Probabilistic Interpretation of Data

A Physicist's Approach

By Guthrie Miller

Acknowledgments

The author gratefully acknowledges stimulating interactions and invaluable assistance during the course of this work from the following individuals.

Bill Inkret, Harry Martz, Tom Waters, Mario Schillaci, Ken Klare, Luiz Bertelli, James Gubernatis, Richard Silver, Peter Groer, Dan Strom, Alan Birchall, John Stather, James Marsh, Matt Puncher, Dunstana Melo, Ray Guilmette, Sergey Romanov, Katya Zaytseva, Vadim Vostrontin, Vladimir Vvedensky, Alan Justus, Mike McNaughton, Dave Wannigman, David Pawel, Iulian Apostoaei, Owen Hoffman, Dick Toohey, Tony James, Thomas Johnson, Matt Pratola, Ion Vasilief, John Klumpp, and Roger Frye.

I am truly grateful, and to those I've overlooked, I apologize. You might as well have been coauthors. Aspects of this work wouldn't have happened without you!

Working with John Klumpp in 2015, the main Fortran code was completely restructured and improved taking advantage of features of modern Fortran. Interactions with John Klumpp and Deepesh Poudel have been very stimulating and helpful.

In this 2017 Glasstree edition, the main change is a complete reworking of Chapter 14 on "Hypothesis Testing".

Supplementary Material

Supplementary material in electronic form can be downloaded from www.ProbabilisticDataInterp.wordpress.com. (Reader comments and questions are welcome.) This material includes answers to all exercises as well as Fortran executables and source code for the various programs discussed in the text. The Fortran source code can be compiled using the the open-source G95 compiler. The .QTI plot files can be opened with QtiPlot from www.qtiplot.com.

Guthrie Miller
Santa Fe, New Mexico
September 2017

Cover art and design by Geoffrey Owen Miller
© 2017, Guthrie Miller

Table of Contents

Chapter 1. Why Probabilistic Data Interpretation?

The title of this little book might very well have been "Bayesian Data Analysis" or "Analysis of Data Using Bayesian Statistics." However, in order to bypass some complexities associated with this terminology (for example, that sometimes "Bayesian Statistics" implies "subjective" probabilities), it was decided to start from the elementary principles of the mathematical description of probabilities. So, in these terms, this book considers the mathematical problem involving conditional probabilities that naturally arises when one analyzes or interprets experimental data.

A physicist's approach to this subject consists in trying to efficiently understand things from first principles and in using a particular style of mathematics that tries to quickly go to the essential ideas, perhaps using examples, without insisting on full mathematical rigor.

Probability always has the meaning of the *frequency distribution* obtained after a large number of trials. These trials are repetitions of "experiments" set up in a well-defined way. The results of these experiments are unpredictable, and such experiments go hand-in-hand with mathematical probability theory. Unpredictability at some level applies to all experimental science, so one might go further and say science and probability go together and that scientific quantities are naturally probabilities. More about that is discussed in the very last chapter of this book. As is essential in science, experiments need to be repeatable, and by following the description of the experiment in the experimenter's lab notebook, the same results can be obtained by anyone. Although the outcome of only one trial of the experiment is irreproducible, the frequency distribution of many trials *is* reproducible.

Probability speaks of events—e.g., a coin toss yields heads—which are produced in some way in a certain experiment, e.g., by tossing a coin. The results or outcomes of the experiment are random variables. The sample space is the set of all possible outcomes of the experiment, which is to say the set of all possible values of the random variable. The probability distribution is the frequency distribution of outcomes in sample space for many repetitions of the experiment.

As an example, consider a coin toss. The outcomes of the experiment are the coin showing either heads or tails. The experiment is that I flip a coin so that it turns over many times in the air, and I catch it in one hand and cover it with the other hand. The covering hand is removed and I look to see whether a head or a tail is showing.

If many repetitions of the experiment are done, in the whole collection of experimental outcomes, there will be about 50% heads and 50% tails; that is, the frequency distribution will be 50% heads and 50% tails. After any coin flip and before showing the coin, there is an unknown true value of the outcome. In this case the chance that the unknown outcome is heads is equal to 50%.

In contrast, consider another experiment. The outcomes are a particular coin lying in my hand showing either heads or tails. The experiment is that I reach into my pocket and produce a coin lying in the palm of my hand. The frequency distribution is 100% heads.

This is because I have learned to detect by touch inside my pocket which side of the coin is heads and turn it heads up before removing it from my pocket.

Therefore, frequency distributions depend on the experiment and not just the sample space.

What if the experiment is not specified? For example, there is a coin in my hand covered by my other hand. It is not meaningful to consider the probability of heads without knowing how the coin got there. If I got the coin from my pocket, depending on what I do before removing the coin, the frequency of heads can be 50%, 100%, or 0%.

However, the coin got there by *some* means. One needs to assume or infer an experiment for the situation to be within the scope of this book. As will be discussed, an initial or prior probability distribution is needed in order to continue the logical progression (e.g., interpretation of a later measurement). One needs to say that there is *some* initial probability distribution (some experiment), even if it is unknown.

In the case of the coin flip there is no uncertain measurement data. One can simply look and see whether the coin is showing heads or tails, and there is no uncertainty in this determination.

To bring in measurement uncertainty, which is the subject of this book, consider an example more relevant to physical science. The experiment is to capture a single atom from a tank containing a large number of "white" and "black" atoms and to measure its color. The measurement of atom color has its own probabilistic nature, and in some fraction of cases a white atom randomly measures black.

The true values of the outcomes of the capture experiment are that the captured atom is white or black. In this case, because atoms are tiny and one cannot see them, the atom color must be measured by another experimental apparatus. After a large number of these measurements, one finds that the frequency distribution of captured atoms is a fraction p white and 1- p black. Because of the large number of these measurements, which causes the measurement uncertainty to average out, this fraction is well known. After a single trial of the atom-capture experiment and before any measurements of atom color, there is an unknown true value of the atom color. The probability that this true value is white is p as obtained from the previous measurements.

The measurement of atom color acts as a filter on the prior probability distribution of atom color. Because of the probabilistic character of the measurement, a measurement result indicating black does not necessarily mean that the atom color is truly black. This is not the only interpretation. The atom might actually be white and the result a false positive.

This prior frequency distribution may not be actually measured, but it is still a real frequency distribution that can be posited and guessed.

Given the sample space, in this case white and black atom color, and the measurement result, one is not able to interpret the measurement data without knowing the experiment that produced the atom that was measured. For example, one experiment is to capture the atom from a tank that contains white and black in a 1 to 1 ratio, and another experiment is to capture the atom from a tank containing a 1000 to 1 ratio. The interpretation of a measurement of atom color will be different in the two cases.

The interpretation is probabilistic; that is, given the measurement result, the *probability* that the true atom color is black is stated. This probability has the usual meaning. Say the probability that the atom color is black is found to be 10%. Then if the experiment of capturing an atom and obtaining the same measurement result were repeated 1000 times, for about 100 of these times the true atom color would be black. In actual practice this process is usually reversed. Some large number, say 1000, of equally probable interpretations of the measurement is calculated using a methodology to be described. If 100 of these interpretations are black, the probability of black is 10%.

In general, the outcome of a measurement is used to infer something about a quantity of interest. Laplace (Laplace 1774) spoke of determining

> *"the probability of the causes of events, a question which has not been given due consideration before, but which deserves even more to be studied, for it is principally from this point of view that the science of chances can be useful in civil life."*

Laplace's cause of an event is the same as the interpretation of the data.

The probabilistic nature of the measurement is encapsulated in a quantity called the likelihood function. It is the probability distribution of measurement values given, or conditioned on, the true value of the quantity of interest. The likelihood function can be determined by other measurements.

In the context of a measurement process, we call the probability distribution of true values of the quantity of interest the "prior." An experimental trial generates one true value of the quantity of interest from this prior distribution, for which a measurement is made. As we shall see, by elementary properties of conditional probability, "Bayes theorem" (a big deal in the time of Bayes and Laplace but something that seems pretty simple now), one can calculate the probability distribution of true values of the quantity of interest (in Laplace's terms, the cause of the

4

measurement event), given the results of the measurement, as the product of the prior probability distribution times the likelihood function. The probability distribution so obtained is called the posterior distribution.

I've tried to present an elementary and uniform approach to this subject. The difficulties are those alluded to early on in the discussion of coin tossing, where an experiment is unspecified, and one knows only the sample space.

An example of this type of problem was considered by Laplace: measurement of the mass of Jupiter. Just as with the coin covered in my hands, there is an unknown true value of the mass of Jupiter. However, the "experiment" that produced Jupiter is unspecified, or perhaps we just don't have the energy and expertise to collect data on similar planets in similar planetary systems. We can say that there is *some* probability distribution of the mass of Jupiter, even if we are not sure what it is. Often, it doesn't make much difference.

Laplace assumed a uniform distribution of the mass of Jupiter for all positive true values. Or, one might also consider a uniform distribution of the log of the mass. Using some measurement technique and then using the measurement result to infer the mass of Jupiter, one arrives at a posterior distribution of the mass of Jupiter based on the assumed prior. For an accurate measurement technique, one finds that the different priors have a very small effect on the posterior distribution.

Even more problematic would be a measurement of the mass of the neutrino. But in this situation also it is possible to imagine a distribution of true values of the neutrino mass coming from a distribution of universes with different laws of physics and testing the sensitivity of the interpretation to the assumptions about this distribution.

The ideas presented in this book are widely understood and accepted as correct, but that acceptance is still not universal. For example, a current international standard (ISO 11929) describes interpretation of measurements without any prior (see exercises 3–7 in Chapter 15), and this is still being done in the high-energy particle physics community (CERN Courier 2002). For sufficiently accurate measurements and sufficiently uniform priors, this may be satisfactory, but otherwise it can lead to important errors. Also, statements about the uncertainty of the result are not straightforward probability statements and are often misinterpreted.

Laplace spoke of the great importance of the science of chances to civil life. This book uses examples from the field of health physics, which very much interfaces with civil life. In the chapter on prior probability distributions the white/black atom example is covered in more detail. It is discussed how the instrument that reads atom color is tested and known to produce false results a small fraction of the time. The overall intent of the measurement is clear—to say something about whether dreaded black atoms are present—and this is well understood by the civil-life client. There is a great difference between just describing the measurement and interpreting, modeling, or inferring the cause of the measurement result. To communicate the result of the measurement without doing a probabilistic interpretation, letting the data "speak for itself," invariably leads to misunderstandings, as illustrated in the cartoon below.

The civil-life client deserves a clear answer about the result of the measurement. He or she deserves an interpretation of the measurement. Because of the probabilistic nature of the measurement process, there is not a single interpretation, as with heads versus tails, for black atoms being present, the mass of Jupiter, or the mass of the neutrino. The main subject of this book is how to calculate many alternate possible interpretations of the measurements. All should be presented to the client. This naturally takes the form of a probability distribution of the quantity of interest, say atom color, the mass of Jupiter, or the mass of the neutrino.

The techniques described in this book utilize Markov Chain Monte Carlo (MCMC) computer calculations. I learned about this truly amazing technique rather late in my career, which started out in experimental high-energy physics. In this book my intent is to present a straightforward elementary approach to MCMC that will allow readers to begin using it for themselves. This method originated in a landmark 1953 paper by Nicholas Metropolis, Arianna and Marshall Rosenbluth, and Augusta and Edward Teller. Rosenbluth, a brilliant physicist then at the beginning of his career, and his wife Arianna "did all the work" (Gubernatis 2005). The results obtained with this method illustrate the great power of simple, rather random searches carried on for a long time, similar, in some ways perhaps, to the evolution of life. This technique will surely become a relied-upon workhorse of scientific data analysis in the future as fast multiprocessor computers become more and more readily available.

References

1. Laplace, Pierre-Simon. "Memoir on the Probability of the Causes of Events." *Statistical Science* 1(3):364–378 (Aug 1986). English translation by S. M. Stigler from original 1774 article.
2. INTERNATIONAL STANDARD ISO 11929, "Determination of the characteristic limits (decision threshold, detection limit and limits of the confidence interval) for measurements of ionizing radiation — Fundamentals and application." First edition, 2010-03-01.
3. CERN Courier 2002, "Physicists and Statisticians Get Technical in Durham." http://cerncourier.com/cws/article/cern/28716.
4. Metropolis, N., A. W. Rosenbluth, M. N. Rosenbluth, A. H. Teller, and E. Teller. "Equation of State Calculation by Fast Computing Machines." *Journal of Chemical Physics* 21(6):1087–1092 (1953).
5. Gubernatis, J., E., "Marshall Rosenbluth and the Metropolis Algorithm", Phys. Plasmas **12** 057303, 5 pages (2005)

Chapter 2. Mathematical Formalism for Probabilities

An experimental result is represented by a random variable. For example, denote by Y a number that represents the outcome of an experiment. The experiment is repeatable and is repeated $i = 1,...n$ times. In one trial of the experiment, the result obtained is Y_i. The average and variance of Y are denoted by

$$\langle Y \rangle \equiv \lim_{N \to \infty} \left(\frac{1}{N} \sum_{i=1}^{N} Y_i \right) \cong \langle Y \rangle_n \equiv \frac{1}{n} \sum_{i=1}^{n} Y_i$$

$$Var(Y) \equiv \left\langle \left(Y - \langle Y \rangle \right)^2 \right\rangle = \left\langle Y^2 - \langle Y \rangle^2 \right\rangle \qquad (1)$$

$$\cong \frac{1}{n-1} \sum_{i}^{n} \left(Y_i - \langle Y \rangle_n \right)^2 = \frac{n}{n-1} \left\langle Y^2 - \langle Y \rangle_n^2 \right\rangle_n$$

The $n-1$ rather than n in the denominator is not important when n is large. Its origin will be discussed later on in Chapter 12. The square root of the variance is called the standard deviation. Note that the standard deviation is the square root of the mean of the square of the deviation from the mean, hence the term "root mean square" that is sometimes used.

The empirical cumulative probability $\Theta_{emp}(Y)$ giving the probability that $Y_i < Y$ is

$$\Theta_{emp}(Y) = \frac{1}{n} \sum_{Y_i < Y} 1 \quad ,$$

that is, it is the fraction of all the data that are less than Y. If the results Y_i are sorted from smallest to largest

$$\Theta_{emp}(Y_i) = \frac{i-1}{n} \cong \frac{i}{n+1} = \theta_i \quad , \qquad (2)$$

with the second form having $n+1$ rather than n in the denominator in order to assign a small nonzero probability to Y being less than the minimum observed value or greater than the maximum observed value. The empirical cumulative probability steps up the same amount at every data point from smallest to largest.

Also, a frequency distribution of Y can be built up, denoted by $P^{(n)}(y)$. This frequency distribution is a histogram of the experimental results, where some number of bins $j = 1,...m$ are defined, and the number of times Y_i falls in the j th bin is recorded.

These constructs are illustrated in Fig. 1.

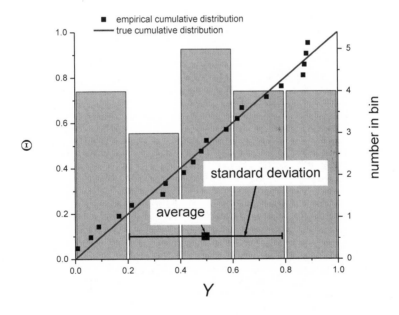

Figure 1—Illustration of basic quantities for 20 repetitions of a random variable Y having a uniform distribution from 0 to 1.

In Figure 1, 20 random numbers, uniformly distributed from 0 to 1, are used as the random variable Y. In this case the true distribution (uniform) is known, and we plot the true cumulative probability along with the empirical cumulative probability from Eq. (2). The histogram is set up with 5 bins covering the interval 0 to 1, and the bar graph with vertical scale on the right-hand side shows the numbers in each bin.

Distribution function. The histogram can be normalized by dividing the number in each bin by the total number of repetitions n. This means that the sum of all the normalized histogram quantities gives 1.

Going, at least conceptually, to the limit $n \rightarrow \infty$, we obtain a nonnegative function $P(Y)$ such that $P(Y)dY$ gives the probability that a result will be in the interval dY and

$$\int_{-\infty}^{\infty} dY\, P(Y) = 1 \ .$$

Obviously this requires a large number of repetitions to be able to have small bins and not-too-small numbers in each bin.

In terms of $P(Y)$, the average, or expectation value, of any function F of Y is defined as

$$Avg(F) \equiv \langle F(Y) \rangle = \int dY\, P(Y)F(Y) \ . \tag{3}$$

Notice that we can also calculate $Avg(F)$ using

$$Avg(F) \cong \frac{1}{n} \sum_{i=1}^{n} F(Y_i) \ , \tag{4}$$

for n large (see exercise 5).

10

The average, variance, and standard deviation are given by

$$Avg(Y) = \langle Y \rangle = \int dY\, P(Y)Y$$

$$Var(Y) = \left\langle \left(Y - \langle Y \rangle\right)^2 \right\rangle = \langle Y^2 \rangle - \langle Y \rangle^2 = \int dY\, P(Y)Y^2 - \left(\int dY\, P(Y)Y\right)^2$$

$$SD = \sqrt{Var(Y)}$$

.

For the uniform distribution shown in Fig. 1, the average is $1/2$, the average of the square is $1/3$, the variance is $1/3 - (1/2)^2 = 1/12$, and the standard deviation is $\sqrt{1/12} \cong 0.289$. The values of the average and standard deviation shown in Fig. 1, obtained for 20 repetitions, are 0.496 and 0.291.

The cumulative probability is given in terms of the distribution function by

$$\Theta(y) = \int_{-\infty}^{y} dY\, P(Y) \quad . \tag{5}$$

In this book the notation Θ or θ is used for probability. Probability is a dimensionless number from 0 to 1. In contrast, a probability distribution function may have dimensions and, although positive, is not limited to be less than 1.

Transformation of one-dimensional distributions. By a change of variables, any one-dimensional probability distribution can be transformed into any other. If it is desired to transform a given distribution P into another distribution Q, then from

$$P(Y)dY = Q(Z)dZ \quad ,$$

we have

$$\Theta_P(Y) = \int_{-\infty}^{y} dY\, P(Y) = \int_{-\infty}^{z} dZ\, Q(Z) = \Theta_Q(Z) \quad ,$$

so that

$$Z(Y) = \Theta_Q^{-1}\big(\Theta_P(Y)\big) \quad .$$

Starting with a distribution P of a variable Y, the transformed variable $Z = \Theta_Q^{-1}\big(\Theta_P(Y)\big)$ has the distribution Q. For example, if we want $Q(Z)$ to be a uniform distribution defined for Z from 0 to 1, and given the fact that

$$\Theta_Q(Z) = Z = \Theta_Q^{-1}(Z) \quad ,$$

we have

$$\begin{aligned}
Z(Y) &= \Theta_P(Y) \\
Y &= \Theta_P^{-1}(Z)
\end{aligned} \qquad . \tag{6}$$

To be consistent with our notational scheme, Z would be denoted by θ.

In modern probability studies, Monte Carlo simulation is a very powerful tool. Computer random number generators generate a random number θ from 0 to 1. If we then apply Eq. (6), the resulting Y's will have the distribution $P(Y)$. This is the way to numerically generate random numbers from *any* one-dimensional distribution.

12

If one has an empirical cumulative probability of the random variable Y_i given by Eq. (2) and wishes to test whether the distribution corresponds to some theoretical distribution P, one can plot $\Theta_P^{-1}\left(\Theta_{emp}(Y_i)\right)$ versus Y_i, which returns Y_i if the distribution is actually P, to see if a straight line is obtained. An example is shown in Fig. 2, where samples are generated from a normal distribution (to be discussed in detail in Chapter 5) using Monte Carlo, and the inverse cumulative probability for the normal is applied.

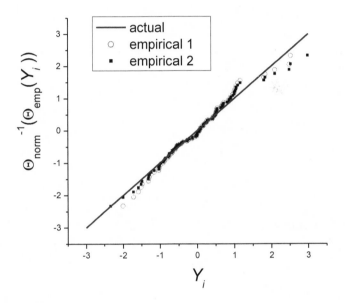

Figure 2—Illustration of testing an empirical cumulative probability (in this case from a normal distribution) against a known distribution using the inverse cumulative probability function of the known distribution applied to the empirical cumulative probability.

The two empirical distributions are that given by the right-hand-side version of Eq. (2) and

$$\Theta(Y_i) = \frac{i-1}{n-1} \quad .$$

By definition, $\Theta(Y_1)$ is the probability that Y is less than Y_1. In the n trials Y_1 is the minimum value, because the values are sorted from minimum to maximum before forming the cumulative probability in this way. Equation (2) assigns a probability $1/(n+1)$ to this missing probability on the left. The alternate cumulative probability assigns zero probability to this missing probability on the left. As one can see from Fig. 2, these details are not very important when the number of samples is fairly large.

Independent variables. Imagine an experiment that involves two random variables, X and Y. These two random variables may be thought of as the two components of a vector quantity. The probability density is then multidimensional, and we have

$$d\Theta = P(X,Y)dX\,dY$$

giving the probability that the vector quantity (X,Y) is in the infinitesimal region $dXdY$.

In terms of the distribution functions, the two random variables X and Y are *independent* if and only if the distribution can be written as the product of two distributions

$$P(X,Y) = Q(X)R(Y) \quad .$$

Conditional probability. For a random variable Y, we define the probability $\Theta(A)$ of the event A to be the probability that the result Y is in some region A of the sample space,

14

$$\Theta(A) = \int_{Y \in A} dY \, P(Y) \quad .$$

The conditional probability $\Theta(A \mid B)$, which is read as the probability of A *given* B, is defined as

$$\Theta(A \mid B) \equiv \frac{\Theta(A \cap B)}{\Theta(B)} \quad , \tag{7}$$

where $A \cap B$ denotes the intersection of the sets A and B, that is, those points in both sets A and B. For a two-dimensional space the intersection is illustrated in Figure 3.

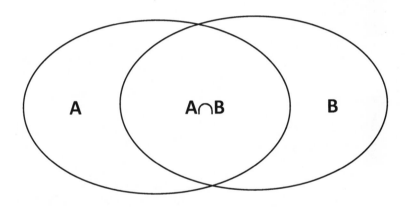

Figure 3—Illustration of the intersection of sets A and B in a two-dimensional space.

Just as in Eq. (7)

$$\Theta(B \mid A) = \frac{\Theta(A \cap B)}{\Theta(A)} \quad ,$$

which immediately proves

$$\Theta(B \mid A) = \frac{\Theta(A \mid B)\Theta(B)}{\Theta(A)} .$$

(8)

Bayes Theorem. In this book, Eq. (8) is used in the following way: There is a random variable Y that is the result of a measurement experiment. The intent of the measurement is to shed light on some model parameter θ, which is also a random variable that comes from a prior distribution $P(\theta)$. The sets A and B are taken as small elements of their respective total spaces: A is a small region ΔY of measurement result space and B is a small region $\Delta\theta$ of parameter space, so that

$$\Theta(A) = P(Y)\Delta Y$$
$$\Theta(B) = P(\theta)\Delta\theta$$
$$\Theta(A \mid B) = P(Y \mid \theta)\Delta Y$$
$$\Theta(B \mid A) = P(\theta \mid Y)\Delta\theta$$

Then it follows from Eq. (8) that

$$P(\theta \mid Y) \propto P(Y \mid \theta)P(\theta) .$$

(9)

In this book, a parameter is often chosen to be a cumulative probability, which explains the θ notation, and in that case the probability density $P(\theta) = 1$ (a constant, which must be 1 by the normalization of $P(\theta)$). This is always possible for a single parameter by a one-dimensional transformation of variables as already explained.

The proportionality relationship of Eq. (9) is simpler than Eq. (8) and more useful in practice because one does not have to bother with the denominator $P(Y)$ explicitly. It is a proportionality relationship with respect to variations of the parameter θ. The denominator comes in naturally as a normalization condition

$$P(\theta \mid Y) = \frac{P(Y \mid \theta)P(\theta)}{\int d\theta \, P(Y \mid \theta)P(\theta)} \quad,$$

if and when it is needed.

In the above relationships we can assume that θ and Y are vector quantities of any dimensionality.

Equation (9) is called Bayes theorem and is of central importance in probabilistic data modeling. It was also derived by Laplace, but he implicitly assumed a uniform prior. Nowadays we see it as an elementary result of the rules of conditional probability.

The intent of the measurement is to shed light on the model parameters. Equation (9) immediately, in one line, gives the answer desired, which is the probability distribution of the parameters given the measurement result.

The quantity $P(Y \mid \theta)$ is called the likelihood function. It, as well as the prior $P(\theta)$, are central to probabilistic data modeling. The likelihood function will be discussed separately in Chapter 7 and the prior in Chapter 8. The prior will be discussed in Chapter 8.

Exercises

1. Using a spreadsheet, generate 20 random numbers as shown in Fig. 1. Hint: Use the built in RAND( ) function.
2. Calculate the average and standard deviation of these numbers. Hint: Use the built in AVERAGE and STDEV functions.
3. Using a spreadsheet, calculate the average and standard deviation of 4, 16, 64, and 256 uniformly distributed random numbers and compare with the expected theoretical result. Look at the dependence of the deviation of the actual result from the expected result as a function of the number of iterations.

4. Using a spreadsheet and Eq. (4), generate 20 random numbers from a normal distribution. Hint: Use the built-in function NORMSINV. Calculate the empirical cumulative probability. Compare this with the calculated theoretical cumulative probability. Hint–Use the built-in function NORMSDIST.

5. Prove the equivalence of Eqs. (3) and (4).

6. Consider the following problem: You examine a large collection of your emails that you regard as spam and notice that the symbol "$" occurs in 90% of these emails. For nonspam emails, the symbol "$" appears only 5% of the time. Also, you see that 5% of your emails are spam. Imagine that you receive 1000 emails. How many do you expect to be in these two categories: spam, nonspam? Now subdivide these, according to whether or not the symbol "$" appears, into these 4 categories: spam-$, spam-no $, nonspam-$, nonspam-no $.

7. In the previous exercise, if the symbol "$" appears in an email, what is the fraction of the number of emails that are spam?

8. In the exercise above, the experiment is that you receive an email. The model parameter is either spam or nonspam. The measurement is that you look to see if the symbol "$" appears. What is the likelihood function giving the probability of measurement results, given the model parameter?

9. Calculate the posterior probability of spam given that you measure $ or no $ using Bayes theorem.

10. What might a person conclude if just the data were quoted in the above exercise? That is, the person is told only that the email contains the $ symbol, and this symbol appears in 90% of spam emails.

11. Show how to calculate the empirical cumulative probability for an integer-valued random variable.

Chapter 3. Binomial, Multinomial, and Poisson Distributions

These distributions are based on conceptual experiments of fundamental importance. We start with the binomial distribution.

Binomial distribution. Consider n trials that can produce two outcomes that we shall call type p and type q having probabilities p and q. The sample space (all possible outcomes) consists of the 2^n terms in the expression

$$1 = (p+q)^n \quad .$$
(1)

For example, after 2 trials,

$$1 = (p+q)^2 = pp + pq + qp + qq \quad .$$

The terms pq and qp both represent the occurrence of p in one trial and of q in another. To show outcomes in terms of the number of p's and q's without regard to their order, we use the binomial expansion of Eq. (1)

$$1 = (p+q)^n = \sum_{k=0}^{n} \frac{n!}{k!(n-k)!} p^k q^{n-k} \quad ,$$
(2)

where the factorial symbol is defined as

$$n! = n \times (n-1) \times ... 2 \times 1 \quad ,$$

so that, for example, $3! = 3 \times 2 \times 1 = 6$.

Each term in the sum in Eq. (2) represents the probability of having k outcomes of type p and $n-k$ outcomes of type q in the n trials.

Multinomial distribution. Similarly, if there are $i = 1,...m$ possible outcomes with probabilities p_i,

$$1 = \left(\sum_{i=1}^{m} p_i \right)^n = n! \sum_{n_i : n = \sum n_i} \prod_{i=1}^{m} \frac{p_i^{n_i}}{n_i!} \quad , \tag{3}$$

where the sum is over all arrangements of numbers of outcomes n_i such that

$$n = \sum_{i=1}^{m} n_i \quad .$$

The distribution given by Eq. (3) is the multinomial distribution.

Poisson distribution. Let us consider in more detail a specific instance where the binomial and multinomial distributions apply. Consider a random variable X. Some number m of histogram bins are set up, and n trials of X are placed in these bins. There is some distribution P of X such that the probability of X being in bin i is given by

$$p_i = \int_{bin\,i} dX\, P(X) \quad .$$

The bins are set up (having a large enough total range to include almost all the probability) so that to a good approximation

$$\sum_{i=1}^{m} p_i = 1$$

The probability of having n_i occurrences of X in bin i is given, from Eq. (2), by

$$P(n_i) = \frac{n!}{n_i!(n-n_i)!} p_i^{n_i} (1-p_i)^{n-n_i} \quad .$$

Now use the approximation

$$(1-p)^{n-k} \cong (1-p)^n \cong e^{-np} \quad ,$$

which apples when $k \ll n$ and p is small. Furthermore,

$$\frac{n!}{(n-k)!} \cong n^k \quad .$$

With these approximations,

$$P(n_i) \cong \frac{\mu_i^{n_i}}{n_i!} e^{-\mu_i} \quad , \tag{4}$$

$$\mu_i \equiv np_i$$

which is the Poisson distribution.

The Poisson approximation applies when the expected number of occurrences in each bin is much less than the total number of trials.

One might also choose to consider the data all together and not bin by bin. From Eq. (3), the probability of having bin occupation numbers $\{n_i\}$ (read $\{.\}$ as "the set of all . ") is then

$$P(\{n_i\}) = n! \prod_{i=1}^{m} \frac{p_i^{n_i}}{n_i!} = \frac{n!}{n^n} \prod_{i=1}^{m} \frac{\mu_i^{n_i}}{n_i!} \quad . \tag{5}$$

$$\mu_i \equiv np_i$$

We expect these two approaches to be consistent. The occurrences in each bin are independent to a good approximation, so we expect

$$P(\{n_i\}) = \prod_{i=1}^{m} P(n_i) \quad ,$$

where $P(n_i)$ is from Eq. (4), and indeed using the fact that

$$\sum_{i=1}^{m} \mu_i = n \quad ,$$

we obtain

$$\frac{\prod_{i=1}^{m} P(n_i)}{P(\{n_i\})} = \frac{n^n e^{-n}}{n!} \quad ,$$

which, aside from being an unimportant constant factor not dependent on $\{n_i\}$, is, for large n, approximately 1 by Stirling's approximation.

As a special case of the above, let the random number X represent the number of disintegrations of n radioactive atoms occurring in a certain time interval. A single atom either disintegrates with probability p or not with probability $1-p$. If the number of atoms is large, the probability of k decays is given by the Poisson formula,

$$P(k) = \frac{\mu^k}{k!} e^{-\mu}$$

$$\mu = np$$

On the other hand, if the number of atoms is not that large and a significant fraction of them decay during the time interval, one must return to the binomial distribution,

$$P(k) = \frac{n!}{k!(n-k)!} p^k (1-p)^{n-k}$$

If there are only two outcomes, say decay or no decay, some simplifications occur in the formulas for the average and variance. Let X be a random variable representing the decay of a single atom. The values of X are either 0 for no decay or 1 for yes decay. Let p be the probability of decay. The distribution function is now discrete, taking only the values $X_k = 0$ and 1. The average and variance of the number of decays for a single atom are given by

$$\langle X \rangle = \sum_{k=0}^{1} X_k P_k = P_1 \equiv p$$

$$Var(X) = \sum_{k=0}^{1} X_k^2 P_k - \langle X \rangle^2 = p(1-p)$$

For a collection of n atoms, because the outcomes for each atom are independent, the average of the sum of the number of decays is the sum of the averages, and the variance is the sum of variances. The mean and standard deviation of the sum of the number of decays are given by

$$mean = np \equiv N$$

$$\frac{SD}{mean} = \frac{\sqrt{1-p}}{\sqrt{N}} \quad .$$

Note that the standard deviation approaches 0 as p approaches 1, because when all the atoms decay, the number of decays is just N without any variation.

As a final example, let the random number X represent the number of cases of a certain observable condition in an epidemiological cohort consisting of N persons. The values of X are either 0 for no effect or 1 for an observed effect, just as in the previous example. For a cohort of N persons, assuming no correlation of the probability of effect from person to person, the above formulas for the mean and standard deviation of the number of effects apply.

Exercises

1. Write out the binomial coefficients for $n = 2, 3, 4$ and 5.
2. A certain disease is known to have a prevalence of 20% in the general population. An extended family group contains 10 people of different ages, etc., similar to the general population. Using a spreadsheet, calculate the probability that the family group experiences 4 or more instances of the disease.
3. Do the above calculation using the Poisson distribution.
4. The disease prevalence in Exercise 2 was obtained from a study population of 100,000, for which 20,000 cases were observed. Using a spreadsheet and assuming a uniform prior, calculate the probability of the disease prevalence given this data.
5. Using a spreadsheet calculate $n^n e^{-n} / n!$ for $n = 30, 60, 120$. What do you observe?

Chapter 4. Example—Poisson Distributed Data

In this chapter, we consider the interpretation of a simple measurement consisting of the detection of some number of counts from radioactive decays.

Counting blank samples. What is being counted (the "samples") are all "blanks"; that is, they are measurements for a sample collected where zero result is expected, or of an empty counting chamber. Thus this measurement is a measurement of background. The situation is simple enough so that we can completely analyze it without a lot of complexity, which facilitates understanding of the basic concept. Furthermore, background measurements are often very important in practice when one wishes to detect something at the lowest possible levels. The problem is then discerning "signal" in the presence of background.

For this example it is assumed at first that 3 background counts have been detected with a uniform prior probability distribution of the true background counting rate λ.

From Bayes theorem with a uniform prior and assuming a Poisson distribution of the number of counts given the true counting rate,

$$P(\lambda \mid N_B) \propto P(N_B \mid \lambda) \propto \lambda^{N_B} e^{-\lambda T} \ . \tag{1}$$

However, this can also be demonstrated rather simply and directly by generating the true background counting rate λ from a uniform prior, then generating the counts N_B from λ, and only considering instances where $N_B = 3$. A Fortran program that does this calculation is shown in Fig. 1.

```
        open(1,file='out.txt') ! output file
        xran=dranr250(1) ! initialize random number generator
        itrial=0
1       continue
        xlambda=dranr250(0) ! uniform distribution from 0 to 1
        nb=ipoisson(xlambda*30) ! Poisson distribution
! in time = 30 mean number of counts = xlambda*30
        if(nb.ne.3) go to 1 ! require that nb = 3
        itrial=itrial+1
        write(1,*)itrial,xlambda !  record xlambda
        if(itrial.lt.10000) go to 1
        end
```

Figure 1—Fortran program that calculates the distribution of true counting rate given a uniform prior and that 3 background counts have been observed. The function ipoisson(μ) returns an integer from a Poisson distribution with mean value μ .

The Monte Carlo–generated (posterior) distribution of true counting rate using this little program is shown in Fig. 2.

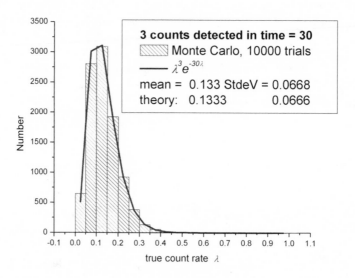

Figure 2—Monte Carlo–generated posterior distribution of true counting rate, given that 3 counts have been detected. As shown, the distribution is a gamma distribution with parameters $\alpha = N_B + 1$, $\beta = T_B$.

The function $\lambda^3 e^{-30\lambda}$ is an example of a gamma distribution, which will be discussed in more detail in Chapter 6. It is defined by

$$g(\lambda \mid \alpha, \beta) = \frac{\beta^{\alpha}}{(\alpha-1)!} \lambda^{\alpha-1} e^{-\beta\lambda} \quad . \tag{2}$$

The normalization integral depends on the definition of the Gamma function,

$$(\alpha-1)! = \Gamma(\alpha) \equiv \int_{0}^{\infty} dx \, x^{\alpha-1} e^{-x} \quad ,$$

where the factorial notation is used.

True counting rate has gamma distribution. We'd like to improve upon the flat prior assumption behind Eq. 1. Let us assume that the true background counting rate λ can be represented by a gamma distribution,

$$P(\lambda)d\lambda = g(\lambda \mid \alpha, \beta)d\lambda \propto \lambda^{\alpha-1} e^{-\beta\lambda} d\lambda \quad , \tag{3}$$

where g is the gamma distribution with parameters α and β.

The gamma parameters α and β can be determined empirically from a large dataset. Using Eq. (3), the probability distribution of measured background counts is the average of the Poisson distribution of counts given true counting rate over the gamma distribution of true counting rate:

$$\begin{aligned} P(N \mid \alpha, \beta) &= \int d\lambda \, g(\lambda \mid \alpha, \beta) \frac{(\lambda T)^{N}}{N!} e^{-\lambda T} \\ &= \left(\frac{R}{R+1}\right)^{\alpha} \left(\frac{1}{R+1}\right)^{N} \frac{(N+\alpha-1)!}{(\alpha-1)! N!} \end{aligned} \tag{4}$$

where $R = \beta / T$ and we have made use of the normalization integral of the gamma distribution.

Assuming in this way that the true counting rate of the background is given by a gamma distribution with some known values of α and β, the distribution of counts N coming from the background is given by Eq. (4). An example is shown in Fig. 3.

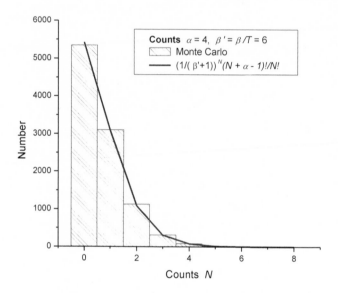

Figure 3—Monte Carlo–generated distribution of counts together with analytical formula, assuming that the background true counting rate is given by a gamma distribution with $\alpha = 4$, $R = \beta / T = 6$.

Such a curve is of practical importance in determining the false positive rate when one is counting samples to detect some radionuclide of interest, if there is a decision level above which the sample is nominally positive. If such a decision level is set, for example at 3 or more counts in Fig. 3, this portion of the distribution, which can be calculated from the given analytical formula, represents the fraction that will be above the decision level just from background, with no radioactivity in the sample. This is the fraction of truly-zero samples that will measure nominally positive, which is the false positive rate.

A useful simplification when α is large, which implies a narrow gamma distribution, and N is not too large is

$$\frac{(N+\alpha-1)!}{(\alpha-1)!} \cong (\alpha-1)^N$$

and Eq. (4) then takes the form of a single Poisson distribution

$$P(N \mid \alpha, \beta) \propto \frac{\mu^N}{N!} \qquad , \tag{5}$$

with

$$\mu = \frac{\alpha-1}{\beta'+1} \cong \frac{\alpha}{R} \qquad .$$

A further numerical simplification is to use a normal approximation of the Poisson distribution with μ the mean of the normal distribution and $\sigma = \sqrt{\mu}$ its standard deviation.

One sees that when the true counting rate comes from a narrow underlying gamma distribution so that the Poisson and normal approximations are valid, one is only able to determine that α and R are large, their ratio, but not how large they are. It is reasonable in this case to limit α to be less than or equal to the total number of background counts for all the background count measurements used to form the distribution.

Empirical determination of gamma distribution parameters. An example of empirically determining α and β' from a large dataset is the following, using 242 measurements of uranium 234 in 24-hour urine "method blank" samples (blank urine carried through the entire analysis procedure). The empirical cumulative probability of background counts is as shown in Fig. 4. Also shown is a fit using the theoretical cumulative probability from Eq. (4). The calculations used the log of the Gamma function to calculate the factorials. The least-squares spreadsheet calculations used to produce Fig. 4 are included in the supplementary material for this Chapter. One can also maximize the combined likelihood function from Eq. (4) to determine α and R or do a full probabilistic analysis (Klumpp et. al. 2014). As discussed in this paper, rebinning of this data reveals that the variation of true counting rate is not purely random but is caused by time variation with a certain correlation time.

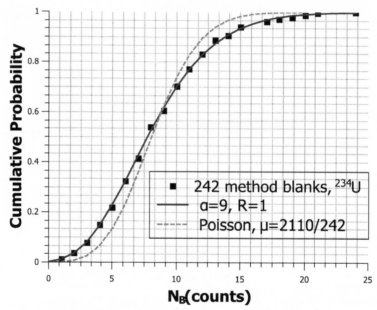

Figure 4—Empirical cumulative probability of background counts of uranium 234 in 24-hour urine for a counting period of 42 hours. The solid curve assumes a gamma distribution of true counting rate with parameters as indicated. The total number of detected counts was 2110 for 242 counting periods.

With such empirically determined $\alpha = \alpha_0$ and $\beta = \beta_0$ for the prior, the interpretation of a single background measurement as in Eq. (1) becomes

$$P(\lambda \mid N_B) \propto P(N_B \mid \lambda) g(\lambda \mid \alpha_0, \beta_0) \propto \lambda^{(N_B + \alpha_0 - 1)} e^{-\lambda(T_B + \beta_0)} \quad,$$

which is again a gamma distribution $g(\lambda \mid N_B + \alpha_0, T_B + \beta_0)$.

That is, the posterior distribution of the counting rate after a measurement of N_B counts detected in a time T_B is given by a gamma distribution with parameters $\alpha = N_B + \alpha_0$ and $\beta = T_B + \beta_0$, where the assumed prior is the gamma distribution $g(\lambda \mid \alpha_0, \beta_0)$ obtained from analysis of other relevant data. One sees particularly clearly in this case how the interpretation of a measurement is influenced by the prior. If the time of the background measurement T_B (and therefore N_B) is increased, eventually the data overwhelm the prior. But the converse is also true. Normally one starts with a uniform prior having $\alpha_0 = 1$ and $\beta_0 \to 0$. In any case the prior is very simple to include in formulas involving α and β by making the substitutions $\alpha = N_B + \alpha_0$ and $\beta = T_B + \beta_0$.

"Moving Target" treatment of background. In many situations a background measurement is not really a measurement of some fixed thing, but a measurement of something thought to be similar to what is contributing the background, for example slight contamination of chemical reagents or slight environmental contamination of the sample. On the other hand, sometimes the background measurement is of a fixed thing, for example contamination of the counting chamber. Both of these situations are handled by making many background measurements and fitting the data to determine the distribution of true background counting rate as illustrated in Fig. 4. Instead of determining the prior probability distribution for a background measurement associated with the sample measurement, an alternative use of this type of data is that the underlying gamma distribution itself is considered to be the distribution of true background counting rate, and there is no single background measurement. In this case $g(\lambda \mid N_B + 1, T_B) \to g(\lambda \mid \alpha, \beta)$, and the background quantities N_B and T_B become

$$N_B = \alpha - 1$$
$$T_B = \beta \qquad ,$$

with α and β determined from a large dataset as done in Fig. 4.

If a fixed background *is* actually being measured, the fitted values of α and β will be large, corresponding to a narrow gamma distribution. In that case as already discussed, it is reasonable to limit α to be less than or equal to the total number of background counts for all the background count measurements used to form the distribution. In the limit of a very narrow gamma distribution, the fitting procedure then gives the same result as one long background count.

Specified counts rather than count time. An alternative to performing successive counts with specified count times is to record the time intervals between counts. For fixed count rate λ, the probability of a time interval t is $d\Theta = e^{-\lambda t}\lambda dt$. Averaging over a gamma distribution $g(\lambda \mid \alpha, \beta)$, the cumulative probability is

$$\Theta(t) = 1 - (1 + t/\beta)^{-\alpha} \quad . \tag{6}$$

Similar to what was done in Fig. 4, a large dataset might be fit to determine the values of α and β, however in the case of a time-interval data, the gamma distribution describes the variation of count rate in a single time interval, while for counts in a specified time period the gamma distribution describes the variation of the mean count rate in the entire period, so the α and β obtained have a different meaning.

Reference

1. Klumpp, J., G. Miller, and A. Brandl. *Characterization of Non-Constant Background in Counting Measurements.* Radiat Prot Dosimetry (2014).

Exercises

1. Derive Eq. (4).
2. Using a spreadsheet, calculate the cumulative probability of the number of counts from Eq. (4). Hint: First calculate the probability distribution for the number of counts and then use the definition of the cumulative probability.
3. Twenty repeated count measurements give the following results: 0, 4, 2, 0, 2, 1, 3, 3, 4, 3, 1, 2, 1, 5, 3, 2, 2, 3, 3, 1. Using a spreadsheet, calculate the best fit values of α and β' in Eq. (4) to this data. Hint: First calculate the empirical cumulative probability. Then, use the SOLVER tool, varying α and β'.

Chapter 5. The Normal Distribution

The normal distribution is defined as

$$P(X \mid X_0, \sigma) = \frac{1}{\sqrt{2\pi}\sigma} \exp\left(-\frac{1}{2}\left(\frac{X - X_0}{\sigma}\right)^2\right) \quad . \tag{1}$$

The quantity X_0 is the mean or average or expectation value. The quantity σ is the standard deviation, which is defined as the square root of the variance.

The normal distribution is one of the most important distributions, because of the central limit theorem, which is discussed here.

It is very helpful to consider the Fourier transform defined by (for any function $f(X)$, not just a positive probability distribution)

$$f_k = \int_{-\infty}^{\infty} dX\, e^{-ikX} f(X) \quad . \tag{2}$$

Fourier transform of normal distribution. The Fourier transform of the normal distribution is calculated by completing the square in the exponent and using the normalization integral. It is given by

$$P_k = \int_{-\infty}^{\infty} dX\, e^{-ikX}\, P(X) = \frac{1}{\sqrt{2\pi}\sigma} \int_{-\infty}^{\infty} dX\, \exp\left\{-\frac{1}{2}\left(\frac{X-X_0}{\sigma}\right)^2 - ikX\right\}$$

$$= \frac{1}{\sqrt{2\pi}\sigma} \int_{-\infty}^{\infty} dX\, \exp\left\{-\frac{1}{2}\left[\left(\frac{X-X_0+ik\sigma^2}{\sigma}\right)^2 + k^2\sigma^2 + 2ikX_0\right]\right\} \qquad (3)$$

$$= \exp\left(-\frac{k^2\sigma^2}{2} - ikX_0\right)$$

The inverse Fourier transform can be obtained using the Fourier representation of the delta function:

$$\delta(X) = \int_{-\infty}^{\infty} \frac{dk}{2\pi} e^{ikX} \quad , \qquad (4)$$

where delta function is the limit of functions with unit integrals that vanish away from zero argument. Therefore,

$$\int \frac{dk}{2\pi} f_k e^{ikX} = \int \frac{dk}{2\pi} dX'\, e^{ik(X-X')} f(X')$$

$$= \int dX'\delta(X-X')f(X') = f(X)\int dX'\,\delta(X-X') = f(X)$$

Similarly to Eq. (4),

$$\delta(k) = \int_{-\infty}^{\infty} \frac{dX}{2\pi} e^{-ikX} \quad . \qquad (5)$$

Fourier transform of convolution. The distribution of two independent random variables is given by the product of the two distributions (independent probabilities multiply). The distribution function of the sum

34

$A = X_1 + X_2$ (adición in Spanish) of two independent random variables X_1 and X_2 with distribution functions P_1 and P_2 is the convolution,

$$P_A(A) = \int dX \, P_1(X)P_2(A - X) \ .$$

The distribution function P_A gives the probability distribution of the sum of the two random variables corresponding to the distribution functions P_1 and P_2.

The Fourier transform of the convolution P_A is given by

$$
\begin{aligned}
P_{A,k} &= \int dX \, e^{-ikX} \, dX' \, P_1(X')P_2(X - X') \\
&= \int dX \, e^{-ikX} \, dX' \frac{dk'}{2\pi} P_{1,k'} e^{ik'X'} \frac{dk''}{2\pi} P_{2,k''} e^{ik''(X-X')} \\
&= \int dX \, e^{-ikX} \frac{dk'}{2\pi} P_{1,k'} \frac{dk''}{2\pi} P_{2,k''} e^{ik''X} \, 2\pi\delta(k' - k'') \\
&= \int dX \, e^{-ikX} \frac{dk'}{2\pi} P_{1,k'} \, P_{2,k'} e^{ik'X} \\
&= P_{1,k} \, P_{2,k}
\end{aligned}
$$

by use of Eq. (5) twice. That is, the Fourier transform of the convolution is the product of the Fourier transforms.

From the Fourier transform of the normal distribution given by Eq. (3) we see that the convolution of two normal distributions is another normal distribution , and for this sum distribution the mean (X_0) is the sum of the two means and the variance (σ^2) is the sum of the two variances.

Central limit theorem. Very often in this book we consider the sums of random variables. If there are $l = 1,...n$ random variables X_l described by the distribution functions P_l, the distribution function of the sum of these random variables

$$A = \sum_{l=1}^{n} X_l$$

is denoted by P_A, and its Fourier transform by

$$P_{A,k} = \prod_{l=1}^{n} P_{l,k} \quad .$$

(6)

Each of the random variables X_l can be redefined to have zero mean by subtracting the mean value $X_{0l} = \langle X_l \rangle$ from X_l. After doing this we represent the Fourier transforms $P_{l,k}$ by a Taylor expansion around $k = 0$:

$$P_{l,k} = \sum_{m=0}^{\infty} \frac{k^m}{m!} \left(\frac{d^m P_{l,k}}{dk^m} \right)_{k=0} \quad .$$

(7)

However going back to the definition of the Fourier transform from Eq. (2),

$$\left(\frac{d^m P_{l,k}}{dk^m} \right)_{k=0} = \int_{-\infty}^{\infty} dX \, (-iX)^m P_l(X) \rightarrow (-i)^m \langle (X_l - X_{0l})^m \rangle \quad ,$$

which gives the moment expansion of the Fourier transform

$$P_{l,k} = \sum_{m=0}^{\infty} \frac{(-ik)^m}{m!} \left\langle (X_l - X_{0l})^m \right\rangle$$

$$= 1 - \frac{k^2}{2} \left\langle (X_l - X_{0l})^2 \right\rangle + \frac{ik^3}{6} \left\langle (X_l - X_{0l})^3 \right\rangle + \ldots \qquad (8)$$

Now we truncate the series in Eq. (8) and consider only the three lowest order terms. Substituting into Eq. (6), we obtain

$$P_{A,k} \cong \prod_{l=1}^{n} (1 - k^2 a_l(k))$$

$$\cong \exp\left(-k^2 \sum_{l=1}^{n} a_l(k) \right) \qquad (9)$$

where

$$a_l(k) = \frac{1}{2} \left\langle (X_l - X_{0l})^2 \right\rangle - \frac{ik}{6} \left\langle (X_l - X_{0l})^3 \right\rangle \qquad (10)$$

To understand the exponential approximation in Eq. (9), start with an exponential approximation of the individual factors, $1 - k^2 a \cong \exp(-k^2 a)$, which is valid for small k^2. The exponential of the sum of terms follows immediately. As k^2 increases, the exponential approximation of individual terms breaks down, but by that time, if n is large, the exponential of the sum is already very small.

Consider the size of the third moment terms relative to the second moment terms in Eq. (9) in the case when all the random variables in the sum are independent and have the same distributions.

$$\frac{3rd\ moment}{2nd\ moment} \approx k \frac{\langle (X - X_0)^3 \rangle}{3\langle (X - X_0)^2 \rangle} \quad .$$

Now notice that the important values of k in the inverse Fourier transform have $k\sigma_n = k\sqrt{n}\sigma$ less than about 1, where

$$\sigma_n^2 \equiv \sum_{l=1}^{n} \langle (X_l - X_{0l})^2 \rangle = n\sigma^2 \quad .$$

The quantity σ_n^2 is the variance of the sum of the n random variables and σ^2 is the average variance. This provides the following bound on the importance of the third moment term

$$\frac{3rd\ moment}{2nd\ moment} < r = \frac{1}{\sqrt{n}} \frac{\langle (X - X_0)^3 \rangle}{\sigma^3} \quad , \tag{11}$$

When the third moment term and other higher order terms are small, the sum distribution has Fourier transform $\exp(-k^2 \sigma_n^2 / 2)$, which corresponds to a normal distribution with variance equal to the sum of variances of the n distributions.

What we have shown is called the central limit theorem, and it is one of the reasons for the importance of the normal distribution. The distribution of the sum of a large number of independent quantities tends to a normal distribution, because from Eq. (11) as $n \to \infty$ the higher order moment terms drop out.

Lognormal distribution. Consider a product of random variables that are all positive. By taking logs, one sees that the distribution of the product of a large number of positive factors tends to a lognormal distribution.

The lognormal distribution is also very important in practice. It is given by

38

$$P_{LN}(X \mid \hat{X}, \sigma)dX \equiv \frac{1}{\sqrt{2\pi} S} \exp\left(-\frac{1}{2}\left(\frac{\log(X/\hat{X})}{S}\right)^2\right) d(\log X)$$

$$\cong \frac{1}{\sqrt{2\pi}\sigma} \exp\left(-\frac{1}{2}\left(\frac{X-\hat{X}}{\sigma}\right)^2\right) dX$$

where the latter approximation comes from

$$\log(X/\hat{X}) = \log\left(\frac{\hat{X} + X - \hat{X}}{\hat{X}}\right) \cong \frac{X - \hat{X}}{\hat{X}}$$

with $\sigma = S\hat{X}$ with $\hat{X}$ denoting the median, which is valid when S is small. In that situation the lognormal distribution is approximately the same as a normal distribution. However, for large S a lognormal becomes distinctively asymmetric with a large tail for positive values. Because $\log X$ has a normal distribution, the mean and the standard deviation of $\log X$ are obviously $\log \hat{X}$ and S.

The mean value of X^p is given by

$$\langle X^p \rangle = \int d(\log X)\, X^p \frac{1}{\sqrt{2\pi} S} \exp\left(-\frac{1}{2}\left(\frac{\log(X/\hat{X})}{S}\right)^2\right)$$

$$= \int d(\log X)\exp(p\log X)\frac{1}{\sqrt{2\pi} S} \exp\left(-\frac{1}{2}\left(\frac{\log(X/\hat{X})}{S}\right)^2\right)$$

which can be evaluated by completing the square in terms of $\log X$.

$$p \log X - \frac{1}{2}\left(\frac{\log(X/\hat{X})}{S}\right)^2 =$$

$$-\frac{1}{2S^2}(\log(X/\hat{X}) - pS^2)^2 + \frac{(pS)^2}{2} + p \log \hat{X}$$

Therefore

$$\langle X^p \rangle = \hat{X}^p \exp\left(\frac{(pS)^2}{2}\right) \quad,$$

and, in particular the mean $X_0 \equiv \langle X \rangle = \hat{X}e^{S^2/2}$, and

$$\langle (X - X_0)^2 \rangle = X_0^2(e^{S^2} - 1)$$

$$\langle (X - X_0)^3 \rangle = X_0^3(e^{S^2} - 1)^2(e^{S^2} + 2)$$

$$\langle (X - X_0)^4 \rangle = X_0^4(e^{S^2} - 1)^2(e^{4S^2} + 2e^{3S^2} + 3e^{2S^2} - 3)$$

(12)

Thus the standard deviation of X is given by

$$\sigma = \sqrt{\langle (X - X_0)^2 \rangle} = X_0\sqrt{e^{S^2} - 1} \quad,$$

where X_0 is the mean.

For S large, the standard deviation divided by the mean is exponentially large. The lognormal is an example of a heavy-tailed distribution where it takes many trials to explore the outlying region. Such distributions are difficult to deal with in practice.

The standard deviation of $\log X$ is S, and sometimes the lognormal is characterized by the geometric standard deviation

$$GSD = \exp(S) \quad,$$

which is the multiplicative factor corresponding to one standard deviation of the log. Figure 1 shows the probability density of three lognormal distributions, with $GSD = 1.6$ $(S = 0.47)$, $GSD = 3$ $(S = 1.1)$, and $GSD = 10$ $(S = 2.3)$. The median $\hat{X}$ in each case is 1.

By solving $dP_{LN}(X)/dX = 0$, one finds that the maximum (mode) of the lognormal in terms of X is given by

$$X_{max} = \hat{X}e^{-S^2} \quad .$$

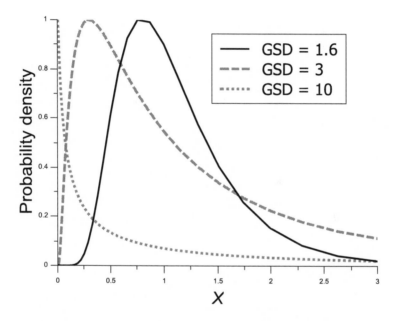

Figure 1—Three lognormal distributions, each having median 1, with successively larger values of the GSD. The normalization is such that the maximum value is 1.

Figure 2 shows the distribution of two sums of lognormally distributed random variables, each having the same value of $r \cong 0.5$ from Eq. (11), one with 300 random variables having $GSD = 3$ and the other with 10 random variables having $GSD = 1.6$. The calculation uses the Fourier transform evaluated using Eqs. (9) and (10).

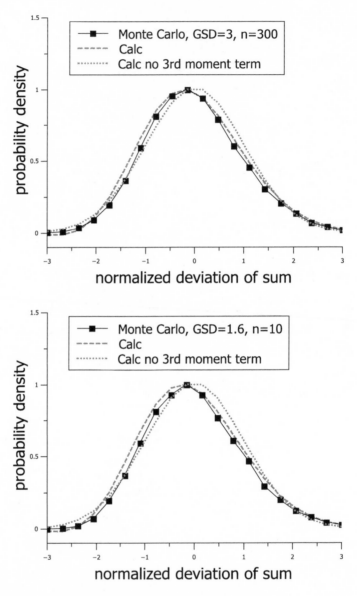

Figure 2—The distributions of two sums of lognormally distributed random variables, both having the same value of the parameter $r \cong 0.5$, which gives an approximate measure of the importance of the third order moment term. The curves are from the numerically calculated Fourier transform using the approximation discussed in the text that takes into account the second and third moments. The data points are from a straightforward Monte Carlo Calculation.

The data points shown in Fig. 2 were generated by Monte Carlo, with 100000 trials, each trial involving the sum of 10 or 300 lognormally distributed random variables. The points in Fig. 2 come from a normalized histogram of these 100000 results. For the cases shown in Fig. 2, the relative size of the third moment from Eq. (11) has the value

$$r = \frac{1}{\sqrt{n}} \frac{\langle (X - X_0)^3 \rangle}{\sigma^3} = \frac{\sqrt{e^{s^2} - 1}(e^{s^2} + 2)}{\sqrt{n}} \cong 0.5 \quad .$$

This type of analysis is useful in considering the approach to the normal limit of the sum of random variables from an asymmetrical distribution.

Median versus average. The central limit theorem tells us that the distribution of the average (mean) of a large number n independent random variables X approaches a normal distribution as $n \rightarrow \infty$ with relative breadth that decreases like $1/\sqrt{n}$. Even if n is too small for central limit theorem to apply, the standard deviation divided by the mean is given by

$$\frac{1}{\sqrt{n}} \frac{\sigma}{X_0} \quad ,$$

where σ and X_0 are the standard deviation and mean of a single variable, because the sum of n independent random variables has mean n times the mean and variance n times the variance (σ^2) of a single variable.

Instead of using the average of the n random variables to obtain a central value , we might also use the median $\hat{X}$ in the hope that this would be less susceptible to outliers and would produce a narrower, more normal distribution. For some asymmetrical distributions this is in fact the case.

The median of n independent random variables approaches a normal distribution centered at the true median with standard deviation of the median

$$\frac{1}{\sqrt{n}} \frac{1}{2P(\hat{X})} \quad ,$$

where $P(\hat{X})$ is the distribution function evaluated at the median (proof by Hironrin, also see Exercise 4). For a lognormal distribution

$$2P(\hat{X}) = \frac{1}{\sqrt{\pi/2}\, S\, \hat{X}} \quad ,$$

and for a lognormal distribution the ratio of breadths, average to median, is given by $\sqrt{e^{s^2}-1}/(S\sqrt{\pi/2})$. A plot is shown in Fig. 3 as a function of GSD ($GSD = \exp(S)$).

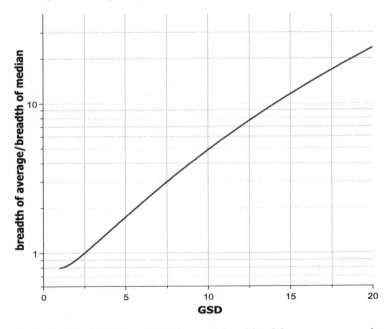

Figure 3—Standard deviation divided by mean (breadth) of the average compared to that for the median of n independent lognormals, showing how the median is tighter than the average for asymmetrical lognormals.

Figure 4 illustrates this for the extreme case shown in Fig. 3, $GSD = 20$. It is remarkable that the median is distributed normally even for this extremely asymmetrical distribution.

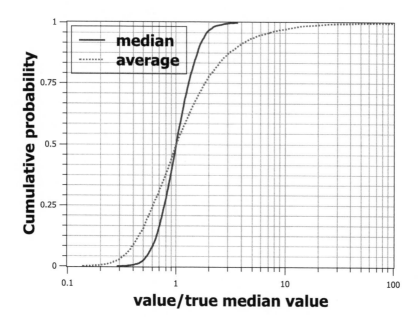

Figure 4—Empirical cumulative probability of the median and of the average of $n = 100$ *random numbers generated from a lognormal distribution with* $GSD = 20$ $(S = 3)$.

Sum of correlated random variables. If the variables are correlated and not independent, the central limit theorem does not hold. For example, if there is 100% correlation, the sum of the average of the random variables has the same distribution as any one.

Consider the general case of a sum of correlated random variables. We denote by A the sum of n, possibly correlated, random variables

$$A = \sum_{i=1}^{n} X_i \quad .$$

The mean and variance of A can be written as

$$\langle A \rangle = \sum_{i=1}^{n} \langle X_i \rangle$$

$$Var(A) = \langle A^2 \rangle - \langle A \rangle^2 = \sum_{i=1}^{n} \sum_{j=1}^{n} (\langle X_i X_j \rangle - \langle X_i \rangle \langle X_j \rangle)$$

By defining the covariance $C_{i,j}$

$$C_{i,j} \equiv \frac{\langle (X_i - \langle X_i \rangle)(X_j - \langle X_j \rangle) \rangle}{\sigma_i \sigma_j}$$

with

$$\sigma_i = \sqrt{\langle (X_i - \langle X_i \rangle)^2 \rangle} \quad ,$$

the variance becomes

$$Var(A) = \sum_{i=1}^{n} \sum_{j=1}^{n} C_{i,j} \sigma_i \sigma_j \quad . \tag{13}$$

The covariance $C_{i,j}$ is the average of the product of two random variables $\Delta X_i / \sigma_i$ and $\Delta X_j / \sigma_j$ both having zero mean and unity standard deviation. The two extreme cases are $\Delta X_i / \sigma_i = \Delta X_j / \sigma_j$ and $\Delta X_i / \sigma_i = -\Delta X_j / \sigma_j$, so that $|C_{i,j}| \leq 1$. It follows from the definition that the covariance is 1 for $i = j$.

To appreciate the meaning of Eq. (13), consider the case where the covariance is zero, except when $i = j$. Then

$$Var(A) = \sum_{i=1}^{n} \sigma_i^2 \quad .$$

However, if there is complete positive correlation so that $C_{i,j}=1$,

$$Var(A) = \left(\sum_{i=1}^{n} \sigma_i \right)^2 .$$

Exercises

1. Using a spreadsheet, generate 50 values from the sum of six uniform distributions. Calculate a normalized deviate taking into account the mean and variance of the sum of six uniform distributions. Calculate the empirical cumulative probability.
2. Now calculate a theoretical normal distribution from the cumulative probability. Hint: Use the built-in function NORMSINV. Compare the deviates with those obtained above.
3. Using a spreadsheet, generate 36 numbers from a lognormal distribution with $GSD = 1.1$. Do this by taking the exponential of the random variable above. Calculate the mean and standard deviation. Hint: Use the built-in functions AVERAGE and STDEV. Compare with the formulas given above. Now do the same for $GSD = 10$. Do the observed and calculated mean and standard deviation still approximately agree? If not, why not? Repeat with only 6 numbers generated in the same way. Using Eq. (11) estimate the number n so that the sum of the n lognormals would appear similar to what is shown in Fig. 2.
4. For a random number x with theoretical cumulative probability $\theta(x)$ show, following Hironrin, that if the empirical cumulative probability curve $\theta_{emp}(x)$ is constructed using a large number n values of x, the distribution of $\theta(x)$ around $\theta_{emp}(x)$ is normal with standard deviation $\sqrt{\theta(1-\theta)/n}$.
5. For a sum of correlated random variables, derive a formula for the variance when the correlation depends only on the difference in sequence number.

Reference

1. Hironrin, Tontoko, www.oocities.org/tontokohirorin/mathematics/clt/median/ctl-median.htm

Chapter 6. The Gamma Distribution

The gamma distribution is defined by

$$g(x \mid \alpha, \beta) \equiv \frac{x^{\alpha-1} e^{-\beta x}}{\int_0^{\infty} dx\, x^{\alpha-1} e^{-\beta x}} = \beta^{\alpha} \frac{x^{\alpha-1} e^{-\beta x}}{\int_0^{\infty} dx\, x^{\alpha-1} e^{-x}} \quad . \tag{1}$$

The normalization integral can be expressed in terms of the the Gamma function,

$$\int_0^{\infty} dx\, x^{\alpha-1} e^{-x} \equiv \Gamma(\alpha) \equiv (\alpha-1)! \quad .$$

We will use the factorial notation for complex as well as integer values of α. Because the defining recursion relationship of the factorial function, $\alpha! = \alpha(\alpha-1)!$, is also satisfied for complex α, it seems least confusing to use the same notation for both.

Note that the gamma distribution can be used to approximate a uniform (flat) distribution by having $\alpha = 1$ and $\beta \to 0$. The limit rather than limiting value $\beta = 0$ is necessary to have a normalizable distribution.

The gamma distribution is closely related to the Poisson distribution and

$$P(N \mid \mu) = \frac{\mu^N e^{-\mu}}{N!} = g(\mu \mid \alpha = N+1, \beta = 1) \quad .$$

Note however that the Poisson distribution is a probability distribution of the integer N , while the gamma distribution is a probability distribution of the continuous variable μ ,

$$\sum_{N=0}^{\infty} P(N \mid \mu) = 1$$

$$\int_{0}^{\infty} d\mu\, g(\mu \mid \alpha, \beta) = 1$$

The peak (mode) of the gamma distribution occurs at the point where the derivative vanishes,

$$\frac{d}{dx}(x^{\alpha-1} e^{-\beta x}) = 0 \quad ,$$

which gives

$$x_0 = \frac{\alpha - 1}{\beta} \quad .$$

If $\alpha \leq 1$ the maximum occurs at $x = 0$. If $x_0 > 0$, by expanding around x_0,

$$x^{\alpha-1} e^{-\beta x} = \exp\big((\alpha - 1)\log(x) - \beta x\big)$$

$$\cong \exp\left((\alpha - 1)\log(x_0) - \beta x_0 - \frac{(\beta(x - x_0))^2}{2(\alpha - 1)} \right) \quad , \tag{2}$$

which reveals the normal approximation to the gamma distribution. Stated in words, the gamma distribution approaches a normal distribution with maximum (mean and mode) at $x_0 = (\alpha - 1)/\beta$ and standard deviation $\sigma = \sqrt{\alpha - 1}/\beta$ as that normal distribution becomes narrow, narrow meaning that

$$S = \frac{\sigma}{x_0} = \frac{1}{\sqrt{\alpha - 1}}$$

is small. In this situation there is also a lognormal approximation of the gamma distribution

$$g(x \mid \alpha, \beta) \propto \exp\left(-\frac{1}{2} \left(\frac{\log(x/x_0)}{S} \right)^2 \right) \quad .$$

The gamma distribution average of x^p is given by

$$Avg(x^p) = \int dx\, x^p\, g(x \mid \alpha, \beta)$$
$$= \int dx\, \frac{\beta^\alpha}{(\alpha-1)!} x^{\alpha+p-1} e^{-\beta x} \quad .$$
$$= \frac{(\alpha+p-1)!}{(\alpha-1)!\,\beta^p}$$

Therefore the average value of x (also referred to as the expectation value or mean value) from the gamma distribution is given by

$$\bar{x} = Avg(x) = \frac{\alpha}{\beta} \quad .$$

The variance is given by

$$Var(x) \equiv Avg\!\left((x-\bar{x})^2\right) = Avg(x^2) - \bar{x}^2$$
$$= \frac{\alpha(\alpha+1)}{\beta^2} - \frac{\alpha^2}{\beta^2} = \frac{\alpha}{\beta^2} \quad .$$

The 3rd and 4th moments are

$$Avg\left((x-\bar{x})^3\right) = \frac{2\alpha}{\beta^3}$$

$$Avg\left((x-\bar{x})^4\right) = \frac{3\alpha(\alpha+2)}{\beta^4} \quad .$$

Convolution property of gamma distribution. The gamma distribution has an important convolution property

$$\int_0^\infty dx'\, g(x-x'\,|\,\alpha_1,\beta)g(x'\,|\alpha_2,\beta) = g(x\,|\,\alpha_1+\alpha_2,\beta) \quad . \tag{3}$$

This can be demonstrated using the Fourier transform defined by

$$f_k = \int_{-\infty}^\infty dx\, e^{-ikx} f(x) \quad . \tag{4}$$

In this case the function $f(x)$ is nonzero only for nonnegative x.

The inverse Fourier transform can be obtained using the Fourier representation of the delta function:

$$\delta(x) = \int_{-\infty}^\infty \frac{dk}{2\pi} e^{ikx} \quad . \tag{5}$$

Similarly to Eq. (5),

$$\delta(k) = \int_{-\infty}^\infty \frac{dx}{2\pi} e^{-ikx} \quad .$$

The Fourier transform of the gamma distribution is given by

$$g_k = \frac{\beta^\alpha}{(\alpha-1)!}\int dx\, e^{-ikx} x^{\alpha-1} e^{-\beta x} = \frac{\beta^\alpha}{(\alpha-1)!}\frac{(\alpha-1)!}{(\beta+ik)^\alpha} = \frac{\beta^\alpha}{(\beta+ik)^\alpha} \quad .$$

We know from Chapter 5 that the Fourier transform of a convolution is the product of the Fourier transforms of the functions that are convoluted. Thus the Fourier transform of the left-hand side of Eq. (3) is given by

$$\frac{\beta^{\alpha_1}}{(\beta+ik)^{\alpha_1}}\frac{\beta^{\alpha_2}}{(\beta+ik)^{\alpha_2}} \quad ,$$

which is the Fourier transform of the Gamma function on the right-hand side of Eq. (3). Thus, the convolution of two gamma distributions with the same β is again a gamma distribution with $\alpha = \alpha_1 + \alpha_2$.

If θ is a random number between 0 and 1, and

$$x = -\log(1-\theta) \quad , \tag{6}$$

then $d\theta = \exp(-x)dx$, so that the distribution of x is $\exp(-x)dx$, which is the gamma distribution for $\alpha = 1$ and $\beta = 1$.

The convolution property of the gamma distribution implies that the gamma distribution for any integer value of α can be built up as the α-fold convolution of the gamma distribution for $\alpha = 1$. This means that the sum of α random numbers of the form given by Eq. (6) has as its distribution the gamma distribution for α and $\beta = 1$. Note also that x generated from gamma (α, β) is the same as y/β, where y is generated from gamma $(\alpha, \beta = 1)$ (see Exercise 1).

52

Fokker-Planck equation. One experiment associated with the gamma distribution is the energy loss experienced by a fast electron passing through matter. The particle energy on exit of a layer of matter will be less than the initial value due to radiation losses. For a given small thickness of the layer, this energy loss is not constant but has considerable fluctuations.

The unknown probability distribution will be denoted by $P(\varepsilon, t)$, where ε is the energy loss and t the thickness of matter traversed. The kinetic equation which defines this function (the "Fokker-Planck" equation) equates the change of the distribution $(\partial P / \partial t)dt$ on a length dt to the "collision integral," which expresses the difference in the number of particles that arrive at, due to radiation losses of higher energy particles along dt, a given total energy loss, and the number of particles that leave the given energy loss interval,

$$\frac{\partial P(\varepsilon, t)}{\partial t} = \int_0^\infty d\varepsilon' \, w(\varepsilon') \big(P(\varepsilon - \varepsilon', t) - P(\varepsilon, t) \big) \quad . \tag{7}$$

The energy loss in an infinitesimal length (caused by the radiation of soft photons) is assumed to have the probability distribution

$$w(\varepsilon)dt = \frac{\alpha \, dt}{\varepsilon} e^{-\beta \varepsilon} \quad , \tag{8}$$

where α is a probability per unit length, and the parameter β limits the distribution for large values of ε. Note that this distribution diverges for small energy loss ε (the so called "infrared divergence").

However, what we are seeking is the probability distribution of total energy loss in a finite interval Δt. We'll see that this probability distribution is given by

$$P(\varepsilon, \Delta t) = \frac{\alpha \, \Delta t}{\varepsilon} (\beta \varepsilon)^{\alpha \Delta t} \frac{\exp(-\beta \varepsilon)}{(\alpha \Delta t)!} \quad . \tag{9}$$

This distribution is a gamma distribution ($\propto \varepsilon^{\alpha \Delta t} \exp(-\beta \varepsilon)$). The function $(\alpha \Delta t)!$ is the Gamma function, given approximately by

$$x! \equiv \Gamma(1+x) \cong 1 - Cx$$

for $x \ll 1$, where C is Euler's constant, $C = 0.577$.

To recognize that Eq. (9) solves Eq. (7), first note that Eq. (8) is $1/\Delta t$ times $P(\varepsilon, \Delta t)$ given by Eq. (9) in the limit $\Delta t \to 0$. We therefore make this substitution in Eq. (7) and go to the limit $\Delta t \to 0$ as the last step. This solves the divergence problem, because Eq. (9) is a normalized probability distribution for all $\Delta t > 0$ no matter how small. Because of the convolution property of the gamma distribution, the first term in Eq. (7) is $P(\varepsilon, t + \Delta t)/\Delta t$ and the second term is $P(\varepsilon, t)/\Delta t$, which together precisely equal the left-hand side after taking the limit $\Delta t \to 0$.

Consider the passage of a high-energy electron through a thickness t of matter, which is imagined to be subdivided into N layers of thickness $\Delta t = t/N$. Because the energy loss in the entire thickness is the sum of energy losses occurring in each of the layers, the distribution of energy loss is the N-fold convolution of energy loss distributions from all the layers, and the average energy is given by N times the expectation energy loss in 1 layer $\varepsilon_1 = \int d\varepsilon\, \varepsilon\, P(\varepsilon, \Delta t) = \alpha \Delta t / \beta$,

$$\langle \varepsilon \rangle = N \varepsilon_1 = \frac{\alpha t}{\beta} \quad .$$

Similarly, the variance of the energy loss in the entire thickness is given by N times the variance of the energy loss in 1 layer, $\sigma_1^2 = \alpha \Delta t / \beta^2$,

$$\sigma^2 \equiv Var(\varepsilon) = N\sigma_1^2 = \langle \varepsilon \rangle \frac{\sigma_1^2}{\varepsilon_1} \quad .$$

By the central limit theorem, the N-fold convolution approaches a normal distribution and

$$P(\varepsilon) \propto \exp\left(-\frac{1}{2}\frac{(\varepsilon - \langle\varepsilon\rangle)^2}{\langle\varepsilon\rangle(\sigma_1^2/\varepsilon_1)}\right) \quad,$$

characterized by a parameter σ_1^2/ε_1, which is the ratio of variance to the mean of the energy loss in a thin layer. This can be evaluated in our model of energy loss using Eq. (9), giving the result

$$\sigma_1^2/\varepsilon_1 = 1/\beta \quad$$

and

$$P(\varepsilon) \propto \exp\left(-\frac{1}{2}\frac{(\beta\varepsilon - \alpha t)^2}{\alpha t}\right) \quad. \tag{10}$$

Exercise 5 discusses the relationship between Eq. (10) and Eq. (9).

Exercises

1. Show that x generated from gamma (α, β) is the same as y/β, where y is generated from gamma $(\alpha, \beta = 1)$.
2. Using a spreadsheet, generate 100 numbers from the gamma distribution $g(x \mid \alpha = 1, \beta)$.
3. Now by using the convolution property of the gamma distribution, generate 100 numbers from $g(x \mid \alpha = 6, \beta)$.
4. Calculate the empirical cumulative probability of the above distribution and compare it to a normal distribution.
5. What is the limit that makes Eq. (9) and Eq. (10) equal?

Chapter 7. The Likelihood Function

The likelihood function is central to the interpretation of data. The likelihood function encapsulates the probabilistic aspects of a measurement. It is the filter that narrows the range of the model parameters interpreting the data.

The measurement is itself an experiment with a random outcome Y and parameter ψ representing the true value of the outcome. The true value of the outcome might be the average or median outcome after many trials. The true value parameterizes the nonrandom aspect of the measurement. The likelihood function is the probability distribution for the data having a measured value Y given ψ, $P(Y \mid \psi)$. In other words, for a given true value ψ, the distribution of measurement results is $P(Y \mid \psi)\, dY$.

In data modeling, the measurement result Y is known and fixed, while ψ is an unknown variable. In data modeling ψ is calculated in terms of other variables θ, which are the parameters of the model. The basic idea is to use the measurement to narrow down the inference about the distribution of model parameters θ.

The likelihood function is defined in terms of a proportionality relationship as

$$L(\psi) \propto P(Y \mid \psi) \quad , \tag{1}$$

considered as a function of ψ. The proportionality means that the likelihood can be freely multiplied by a constant with respect to ψ. We choose the normalization of the likelihood function for a single measurement such that the maximum is 1. Normalized in this way, the likelihood function for a single measurement can be written as

$$L(\psi) = \exp\left(-\frac{1}{2}\chi(\psi)^2\right) \quad,$$

where

$$\chi(\psi) = \pm\sqrt{-2\log(L(\psi))} \quad.$$

The quantity $\chi(\psi)$ is the number of standard deviations the measurement deviates from the predicted true value from the model calculation $\psi = \psi(\theta)$.

If there are multiple independent measurements, the combined likelihood function for all the data is the product of the likelihood functions for the individual measurements (see Exercise 1).

Exact Poisson likelihood. As we have already seen, the Poisson likelihood is given by a gamma distribution,

$$L(\mu) \propto \mu^N e^{-\mu} \quad,$$

where the data consist of the number of detected counts N and the parameter μ is the true mean number of counts. Written in normalized form,

$$L(\mu) = \frac{\mu^N e^{-\mu}}{N^N e^{-N}} = \exp(N\log(\mu/N) + N - \mu) = \exp\left(-\frac{1}{2}\chi(\mu)^2\right).$$

$$\chi(\mu) = \pm\sqrt{2(\mu - N - N\log(\mu/N)}$$

When the counts are large, the normal or lognormal approximations may be used. In the normal approximation, by expanding the log term to second order for small $\mu - N$,

$$L(\mu) \cong \exp\left(-\frac{1}{2}\left(\frac{\mu - N}{\sigma}\right)^2\right) \quad,$$

where the uncertainty standard deviation of the number of counts $\sigma = \sqrt{N}$. Alternatively, in the lognormal approximation, using the formula $x_1 - x_2 \cong x_2 \log(x_2 / x_1)$, which applies for x_1 near x_2,

$$\mu - N \cong -N \log\left(\frac{\mu}{N}\right) \quad,$$

which leads to

$$L(\mu) \cong \exp\left(-\frac{1}{2}\left(\frac{\log(\mu / N))}{S}\right)^2\right) \quad,$$

where $S = \sigma / N = 1 / \sqrt{N}$.

Counting measurement with background. We now consider a counting measurement more carefully, starting with the changes required when there is a background subtraction, which is almost always the case in practice.

The true value of the mean number of counts μ is assumed to be related to the true value of the measured quantity ψ by a normalizing coefficient A and a background B,

$$\mu = A\psi + B \quad, \tag{2}$$

where the normalizing coefficient A is first assumed to be a constant. Later we will consider variability of A.

A background measurement has been discussed in Chapter 4, and following that discussion $B = \lambda_B T$, where λ_B is the background counting rate, which is given by a gamma distribution determined by previous measurement of the background, and T is the sample count time.

The likelihood function, when N counts are detected, is then given by

$$L(\psi) \propto \int d\lambda_B \left(A\psi + \lambda_B T\right)^N e^{-(A\psi + \lambda_B T)} g(\lambda_B | \alpha_B, \beta_B) \quad , \qquad (3)$$

where the background counting rate, summarized by its posterior distribution, comes from a gamma distribution as discussed in Chapter 4. For a single background measurement detecting N_B counts in counting period T_B and assuming a gamma (α_0, β_0) prior on background counting rate, $\alpha_B = N_B + \alpha_0$, and $\beta_B = T_B + \beta_0$. For a noninformative background prior, $\alpha_0 = 1$ and $\beta_0 \to 0$. Alternately, as described in Chapter 4, α_B and β_B can be obtained from a large-dataset background study.

The integral in Eq. (3) can be calculated using numerical integration or using Monte Carlo, by generating λ_B from its gamma distribution and summing,

$$L(\psi) \propto \frac{1}{N_j} \sum_j (A\psi + \lambda_B^{(j)} T)^N \exp\left(-(A\psi + \lambda_B^{(j)} T)\right) \quad , \qquad (4)$$

where N_j is the number of trials of $\lambda_B^{(j)}$ from its gamma distribution.

This likelihood function is the simplest version of what is termed the "exact" likelihood, which is meant to include all real complexities of the measurement. The exact likelihood function can be calculated numerically for each measurement for a range of values of ψ and stored as an interpolation table. So, whereas one is familiar with having each measurement summarized by a central value and a standard deviation, now each measurement is summarized by a whole table of numbers.

The exact likelihood can also be investigated analytically. If we replace the distributions in Eq. (3) by their normal approximations, we have

$$L(\psi) \propto \int_0^\infty d\lambda_B \exp\left\{-\frac{1}{2}\left[\frac{(N - A\psi - \lambda_B T)^2}{N} + \frac{(\lambda_B T - (\alpha_B - 1)/R)^2}{(\alpha_B - 1)/R^2}\right]\right\}$$

where

$$R \equiv \beta_B / T \quad .$$

The integral can be evaluated analytically as a convolution of two normals, if the lower limit of integration is replaced by $-\infty$, which is permissible if α_B is large. The convolution theorem for normal distributions of the form

$$N(u \mid x, \sigma) = \frac{1}{\sqrt{2\pi}\sigma} \exp\left[-\frac{1}{2}\left(\frac{u - x}{\sigma}\right)^2\right]$$

can be stated as

$$\int_{-\infty}^\infty dv \, N(u - v \mid x_1, \sigma_1) N(v \mid x_2, \sigma_2) = N\left(u \mid x_1 + x_2, \sqrt{\sigma_1^2 + \sigma_2^2}\right) \quad . \qquad (5)$$

If the variables in the theorem are $u = -A\psi$, $v = \lambda_B T$, $x_1 = -N$, $x_2 = (\alpha_B - 1)/R$, $\sigma_1^2 = N$, and $\sigma_2^2 = (\alpha_B - 1)/R^2$, after the convolution integral over $d\lambda_B T$, the resulting normal has center at $A\psi = -(x_1 + x_2) = N - (\alpha_B - 1)/R$, and standard deviation $\sigma = \sqrt{N + (\alpha_B - 1)/R^2}$. Optionally, one can approximate the normal by a lognormal.

Thus, in summary, when the counting uncertainty is small, we have a normal approximation

$$L(\psi) \propto \exp\left(-\frac{1}{2}\left(\frac{\psi - y_{meas}}{\sigma_{meas}}\right)^2\right) \quad , \qquad (6)$$

or a lognormal approximation

$$L(\psi) \propto \exp\left(-\frac{1}{2}\left(\frac{\log(\psi / y_{meas})}{S_{meas}}\right)^2\right) \quad , \qquad (7)$$

60

with

$$y_{meas} = \left(N - (\alpha_B - 1)/R\right)/A$$

$$\sigma_{meas} = \sqrt{N + \alpha_B/R^2}/A \qquad .$$

$$S_{meas} = \sigma_{meas}/y_{meas}$$

Note the substitution of α_B for $\alpha_B - 1$ in the formula for the standard deviation. When α_B is large, α_B is the same as $\alpha_B - 1$, but the formula given seems better from numerical studies, and it prevents $\sigma_{meas} = 0$ for 0 counts and 0 background counts counts, where $\alpha_B = 1$, which is the value that corresponds to a flat background prior.

The lognormal approximation from Eq. (7) is useful and accurate when the counts are large; however, being lognormal, it implies that the probability of $\psi = 0$ is 0 whether or not this is true, so this form is completely inadequate when one wants to examine the question of whether the true value of the measurement ψ might be 0. The normal approximation gives a nonzero result for the probability of $\psi = 0$, but the approximation breaks down in the cases when this probability is significant, when the counts are small. It is often an important problem in practice to determine the probability of $\psi = 0$, for example, determining whether some nonzero amount of a contaminant of interest is present in a sample or that the data are consistent with zero true amount.

To evaluate Eq. (3) without approximation, we use the binominal expansion,

$$(A\psi + \lambda_B T)^N = \sum_{k=0}^{N} \frac{N!}{k!(N-k)!} (A\psi)^k (\lambda_B T)^{N-k} \qquad .$$

The likelihood function from Eq. (3) is therefore

$$L(\psi) \propto e^{-A\psi} \sum_{k=0}^{N} (A\psi)^k \frac{N!}{k!(N-k)!} \int d\lambda_B \, (\lambda_B T)^{N-k} \lambda_B^{\alpha_B - 1} e^{-\lambda_B T(1+R)}$$

Now using the normalization integral of the gamma distribution,

$$\int d\lambda_B \ \lambda_B^{N-k+\alpha_B-1} e^{-\lambda_B T(T+R)} = \frac{(N-k+\alpha_B-1)!}{(T(1+R))^{N-k+\alpha_B}} \quad ,$$

the likelihood function is given by

$$L(\psi) \propto e^{-A\psi} \sum_{k=0}^{N} (A\psi)^k \ \frac{N!}{k!(N-k)!} \ \frac{T^{N-k}(N-k+\alpha_B-1)!}{(T(1+R))^{N-k+\alpha_B}}$$

$$\propto e^{-A\psi} \sum_{k=0}^{N} (A\psi(1+R))^k \ \frac{(N-k+\alpha_B-1)!}{k!(N-k)!} \tag{8}$$

Because of background subtraction, the likelihood function, instead of being a single gamma distribution, is a mixture of $N+1$ gamma distributions, where N is the number of detected sample counts.

A comparison of the exact result using Eq. (8) with Monte Carlo integration using Eq. (3) and the approximate formulas given by Eqs. (6) and (7) for $\alpha_B = 7$, $R = 6$ is shown in Fig. 1.

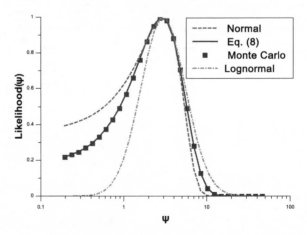

Figure 1—Exact likelihood for 4 sample counts and 6 background counts ($\alpha_B = 7$) in 6 sample counting periods compared with Monte Carlo calculation and normal and lognormal approximations.

In Fig. 1, the result from Eq. (8) and the Monte Carlo integration agree very well, as they must. The normal approximation is qualitatively similar to the correct result, and the lognormal approximation of course gives zero probability as $\psi \rightarrow 0$. In our use of the likelihood function later on to discern whether the measurement is consistent with zero, we will see that one of the most important features of the likelihood function to get right is the ratio of the peak area to the value for $\psi \rightarrow 0$. The normal approximation approximates the peak area but does not calculate $\psi \rightarrow 0$ accurately, so it should not be used in this situation. Rather, the exact likelihood function should be used instead.

Variability of the normalizing coefficient. Let us now consider variability of the normalizing coefficient A appearing in Eq. (2). Imagine that the measurement is of the number of radioactive decays from a urine sample after it has gone through chemical processing. The true value of the measured quantity is the 24h urine excretion after intake of a radionuclide. The coefficient A relates the true value of the measured quantity to the number of counts detected. It is the product of several factors. A urine sample is collected, chemical separation of the element of concern is performed, and the separated element deposited on a counting substrate. Finally, a detector is positioned close to the substrate and the decays of the radionuclide of concern are counted for a time Δt_c. The normalizing coefficient is given by a product of terms for each of these processes

$$A = \Delta t_x \varepsilon_r \varepsilon_c \Delta t_c \quad .$$

The urine sample represents excretion for some time Δt_x (in units of days), the efficiency of the chemical separation (chemical recovery fraction) is ε_r, and the counting efficiency is ε_c. This normalizing coefficient A is quite variable, in this case mostly because of the biological variability of urine excretion, that is, the variability of Δt_x. This variability depends on the urine collection protocol; for example, careful true-24h collection, a creatinine-normalized sample, a specific-gravity normalized sample, a volume-normalized spot sample of some small volume, etc. Studies have been done of this uncertainty by collecting a large number of samples where the true excretion is constant and large enough so that it is measurable with small counting uncertainty. These studies have shown that the variability is approximately lognormal with lognormal standard deviation S from about 0.1 to about 0.6 in the progression above, with the spot sample having the largest variability. The other factors are modeled as lognormal also, so that the overall logarithmic variability of the normalization is given by

$$S_{norm} = \sqrt{\sum_i S_i^2} \quad ,$$

where S_i is the logarithmic standard deviation of the i th factor.

Notice that the variability associated with the normalization is naturally lognormal, because the normalization coefficient is the product of a number of factors, and the product of a large number of quantities becomes lognormally distributed by the central limit theorem.

We also assume that the background B is uncorrelated with the normalizing coefficient A. If the background is caused by a background level of the radionuclide being excreted by the subject, the background would share the factor A and therefore be correlated with A. For example, biological variability would be shared between normalization coefficient and background, because the background is excreted by the person. In this case Eq. (8) is no longer valid, and one must return to the Monte Carlo calculation of the exact likelihood function given by Eq. (4), but with variable A, where this correlation can be simulated.

With no background correlation the exact likelihood including variability of the normalization is given by

$$L(\psi) \propto \int dA P_A(A) e^{-A\psi} N! \sum_{k=0}^{N} \frac{(N-k+\alpha_B-1)!}{k!(N-k)!} \left(A\psi(1+R)\right)^k \quad , \quad (9)$$

where $P_A(A)$ is the probability distribution of the normalization coefficient. Equation (9) can be rapidly computed using a good one-dimensional integration routine. When the counts are large, the normal

64

approximation Eq. (6) can be substituted for the right-hand side terms in Eq. (9), and the normal-counting-uncertainty version of Eq. (9) is

$$L(\psi) \propto \int dA P_A(A) \; \exp\left\{-\frac{1}{2}\left(\frac{\psi A / A_0 - y_{meas}}{\sigma_{meas}}\right)^2\right\} \quad .$$

In the limit of small counting uncertainty, the Gaussian can be replaced by the delta function $\delta(A - A_0 y_{meas} / \psi)\sqrt{2\pi}\sigma_{meas} A_0 / \psi$, yielding

$$L(\psi) \propto P_A(A_0 y_{meas} / \psi) / \psi \quad ,$$

which shows how the distribution of the normalization coefficient completely determines the likelihood function in this limit.

For a lognormal distribution of the normalization coefficient with σ_{meas} / y_{meas} small,

$$dA P_A(A) \propto \frac{dA}{A}\exp\left(-\frac{1}{2}\left(\frac{\log(A/A_0)}{S_{norm}}\right)^2\right) \quad .$$

We can make the logarithmic substitution $[x_1 - x_2 \cong x_2 \log(x_2 / x_1)]$

$$\left(\frac{\psi A / A_0 - y_{meas}}{\sigma_{meas}}\right) \cong \left(\frac{\log(y_{meas} / \psi) - \log(A/A_0)}{\sigma_{meas} / y_{meas}}\right) \quad ,$$

and the likelihood function can then be evaluated as a convolution integral. In the convolution theorem, Eq. (5), $u = \log(y_{meas} / \psi)$, $v = \log(A/A_0)$, and $x_1 = x_2 = 0$. This gives

$$L(\psi) \propto \exp\left(-\frac{1}{2}\left(\frac{\log(y_{meas} / \psi)}{S}\right)^2\right)$$

$$S = \sqrt{S_{norm}^2 + (\sigma_{meas} / y_{meas})^2} \quad , \tag{10}$$

$$y_{meas} = \left(N - (\alpha_B - 1)/R\right)/A_0$$

$$\sigma_{meas} = \sqrt{N + \alpha_B / R^2} / A_0$$

where A_0 is the central (median) value of A.

Figure 2 shows the likelihood function for $N = 10$, $\alpha_B = 4$, and $R = 6.43$, which gives $y_{meas} / \sigma_{meas} = 3$ (a measurement 3 standard deviations positive). The "normal/lognormal" case uses the normal approximation version of Eq. (9) with lognormal distribution of the normalization, and the "exact" case uses Eq. (9). The large difference in the value of the likelihood function at $\psi = 0$ is very significant, and argues strongly for the use of the exact likelihood whenever count quantities are available.

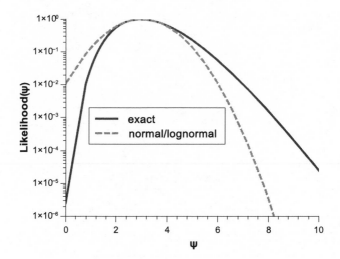

Figure 2—Likelihood function for $N = 10$, $\alpha_B = 6$ ($N_B = 5$), and $R = 6.43$, which gives a measurement result 3 standard deviations greater than 0. The exact Poisson likelihood function and the normal/lognormal form are shown.

When the counts are very large, use of the exact likelihood leads to numerical difficulties, but it is permissible in that case to replace Eq. (9) by its normal/lognormal approximation version, because the details of likelihood function away from its maximum value are then unimportant.

From Eq. (9) when no sample counts are detected,

$$L(\psi) \propto \int dA P_A(A) e^{-A\psi} \quad ,$$

independently of the number of background counts, which shows that for zero sample counts, the maximum of the likelihood occurs at $\psi = 0$.

Likelihood function for "less than" measurement. In the past (hopefully not going forward), measurement results were sometimes recorded as "less than" a certain limit, rather than recording the actual result. This is a very bad practice, because it reduces the capability for low-level detection.

Say this limit is N counts. Then, instead of Eq. (4), the Monte Carlo integration to obtain the likelihood function is given by

$$P(\psi \mid N) \propto \frac{1}{N_j} \sum_{j=1}^{N_j} \sum_{n=0}^{N-1} \frac{(A\psi + \lambda_j T)^n}{n!} \exp\left(-(A\psi + \lambda_j T)\right) \quad ,$$

where the background counting rate λ_j is generated from its gamma distribution for N_j trials. The variability of the normalization coefficient A can also be included by summing also over trials of A from a lognormal distribution. The likelihood function is then as shown in Fig. 3.

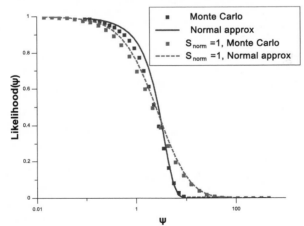

Figure 3—Likelihood function for a "less then" measurement. The decision level is 4 sample counts with $\alpha_B = 10$ and $R = 6$.

The normal approximation shown in Fig. 3 is given by

$$L(\psi) = 1 - \Theta_{norm}\left(\frac{\psi - y_{lim}}{\sigma_0}\right) \quad ,$$

where Θ_{norm} is the normal cumulative probability, σ_0 is the standard deviation for zero result in the normal approximation and $y_{lim} = k_\alpha \sigma_0$ (see Exercise 5). The detection limit N in counts corresponds to y_{lim} in the normal approximation.

General form of the likelihood function. We have chosen a normalization of the likelihood function for a single measurement so that the maximum value is 1. As we have seen, this normalization allows the general representation

$$L(\psi) = \exp\left(-\frac{1}{2}\chi(\psi)^2\right) \quad .$$

As an example, let us return to the Poisson distribution. A plot of $\chi(\psi)$ for different numbers of counts N is shown in Fig. 4.

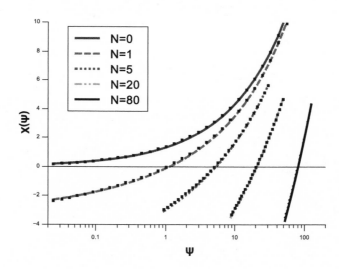

Figure 4—The function giving the scaled residual or normal deviate for Poisson data.

Figure 5 shows the same sequence of gross counts but this time with $\alpha_B = 4$, background scaling factor $R = 1$ and lognormal normalization uncertainty $S = 0.3$.

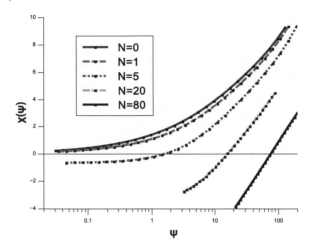

Figure 5—Exact likelihood function normal deviate with gross counts as shown and
$\alpha_B = 4$ ($N_B = 3$), *background scaling factor* $R = 1$, *and lognormal*
normalization uncertainty $S_{norm} = 0.3$.

The general nonlinear problem of maximizing the combined exact likelihood is equivalent to the problem of minimizing the square of the scaled residual, that is, it is a nonlinear least-squares minimization.

The exact likelihood function is summarized by a table of the "scaled residuals" or normal deviates $\chi(\psi)$ as a function of ψ. The use of an interpolation table giving $\chi(\psi)$ rather than one giving $L(\psi)$ allows more accurate extrapolation, which is important for Markov Chain Monte Carlo, where the chain starts out very far from the mode of the combined likelihood function. The quantity $\chi(0)$ has a nice interpretation as the "number of standard deviations" the data point is positive.

Empirical determination of the likelihood function. Imagine $i = 1..., N$ repeated independent measurements $Y^{(i)}$ of a quantity Y having average value $\langle Y \rangle_N = \sum Y^{(i)} / N$. We wish to obtain the combined likelihood function for Y just from the data alone. The true value of Y is denoted by ψ. The combined likelihood function is the product of the N individual likelihood functions, and for the logarithm we may write

$$-2\ln(L(\psi)) = const + \sum_{i=1}^{N} \chi^2(\psi, Y^{(i)}) \ . \tag{11}$$

We Taylor expand Eq. (11) around the point ψ_0, $\langle Y \rangle_N$, where $\chi(\psi_0, \langle Y \rangle_N) = 0$. The expansion variables are $x = \psi - \psi_0$ and $y_i = Y^{(i)} - \langle Y \rangle_N$. To 2nd order in smallness in both,

$$-2\ln(L(\psi)) = const + N\left(Ax^2 + (Bx + Cx^2)\frac{1}{N}\sum_{i=1}^{N} y_i^2 \right) + \dots \ ,$$

for some coefficients $A = \left(\dfrac{\partial \chi}{\partial \psi}(\psi_0, \langle Y \rangle_N) \right)^2$, B and C given by similar expressions involving products of partial derivatives (see Exercise 7). Terms proportional to y_i vanish in the summation over i. Higher order terms are unimportant if x is small.

As a result of combining all the data, this quadratic in x has a minimum shifted away from $x = 0$. As $N \to \infty$, this leading order expression produces large changes in the log likelihood causing the likelihood function to be exponentially small unless x is small, in which case higher order terms can be neglected. Thus the combined likelihood approaches a normal.

What are the parameters of this normal? Two approaches are offered, the conventional one and a more conservative one.

Conventional approach. The empirical combined likelihood function is usually constructed using the average and standard deviation of the data, which are identified with the parameters of the normal likelihood function, without taking into account the large uncertainty of the standard deviation determined from a small number of data. Without a log transformation, the formulas are:

$$\langle Y \rangle_N = \frac{1}{N} \sum_{i=1}^{N} Y^{(i)}$$

$$\langle Y^2 \rangle_N = \frac{1}{N} \sum_{i=1}^{N} \left(Y^{(i)} \right)^2 \qquad , \qquad (12)$$

$$s_Y^2 = \frac{1}{N} \sum_{i=1}^{N} \left(Y^{(i)} - \langle Y \rangle_N \right)^2 = \langle Y^2 \rangle_N - \langle Y \rangle_N^2$$

where s_Y denotes the standard deviation of the data. As seems more natural in this and the next section, the variance (s_Y^2) for a finite sample is defined with N in the denominator rather than $N-1$ as in Eq. (1) of Chapter 2.

The lognormal likelihood function using log-transformed quantities is given by

$$L(\psi) = \exp\left(-\frac{N}{2} \left(\frac{\log \psi - \langle \log Y \rangle_N}{s_{\log Y}} \right)^2 \right) \qquad , \qquad (13)$$

where $s_{\log Y}$ is the standard deviation of $\log Y$ calculated using Eq. (12) from the sample of N values.

If some of the data are zero or negative, the lognormal distribution cannot be used and we would instead use the normal approximation

$$L(\psi) = \exp\left(-\frac{N}{2} \left(\frac{\psi - \langle Y \rangle_N}{s_Y} \right)^2 \right) \qquad ,$$

with the average and standard deviation obtained from Eq. (12).

A more conservative empirical likelihood function. Just as in the previous section the data consist of some number $i = 1..., N$ of independent measurements giving results $Y^{(i)}$. The likelihood function of a single measurement is assumed to be Gaussian,

$$P(Y \mid \psi, \sigma) = \frac{1}{\sqrt{2\pi}\sigma} \exp\left(-\frac{1}{2} \left(\frac{\psi - Y}{\sigma} \right)^2 \right) \qquad ,$$

however with unknown mean ψ and unknown standard deviation σ. Therefore the combined likelihood function is given by

$$
\begin{aligned}
L(\psi,\sigma) &\propto \sigma^{-N}\exp\left(-\frac{1}{2}\sum_{i=1}^{N}\left(\frac{\psi-Y^{(i)}}{\sigma}\right)^{2}\right) \\
&\propto \sigma^{-N}\exp\left(-N\frac{s_Y^2}{2\sigma^2}\left(1+\frac{\left(\psi-\langle Y\rangle_N\right)^2}{s_Y^2}\right)\right)
\end{aligned}
\tag{14}
$$

where, just by doing some algebra, one sees that the data enter in only through the mean and standard deviation (from the sample of N values) given by Eq. (12).

In the calculation in this section, the value of σ is allowed to emerge probabilistically from the data. Considered as a function of ψ, the likelihood function given by Eq. (14) is broad when σ is large and narrow when σ is small. However, the distribution of σ itself also comes from Eq. (14), and this distribution becomes narrower as more data are taken. The likelihood function we are seeking is Eq. (14) multiplied by a prior probability distribution $P(\sigma)$, as will be discussed in Chapter 8, and integrated over all possible values of σ. As will be discussed in Chapter 8, by scale invariance, the natural prior on σ is log-scale uniform in σ, so that $P(\sigma)\propto d\log(\sigma)=d\sigma/\sigma$. Alternatively, a prior that imposes a finite maximum on σ might be used. In either case the integral over σ converges (see Exercise 8) and gives, using the log-scale uniform prior on σ and for log-transformed data, the result

$$
L(\psi)\propto \int L(\psi,\sigma)P(\sigma)d\sigma \propto \left(1+\frac{\left(\log\psi-\langle\log Y\rangle_N\right)^2}{NS^2}\right)^{-N/2},
\tag{15}
$$

where $S=\sqrt{s_{\log Y}^2/N}$ is the standard deviation obtained from the N measurements of the average value $\langle\log Y\rangle_N$. Without a log transformation a similar formula applies. When N is large this likelihood function is normal (as a function of $\log\psi$) of the form given by Eq. (13).

When N is small, the likelihood function given by Eq. (15) is much broader than a normal as shown in Fig. 7.

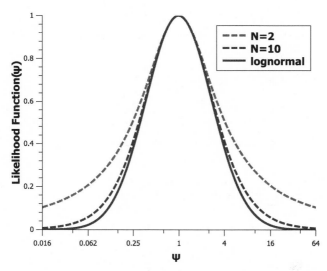

Figure 7—Combined likelihood function for N measurements discussed in this section compared to a lognormal. The average of the log of the data is 0, and the standard deviation of that average (S) is 1 .

Likelihood function for nonindependent Gaussian data. The data consist of a vector with components denoted by Y_k. The data are correlated so that there are off-diagonal terms in the averages of deviations from the means

$$\langle (Y_k - \psi_k)(Y_l - \psi_l) \rangle = \beta_{k,l} \quad , \tag{16}$$

with ψ_k the true mean value of Y_k. It is assumed that the covariance matrix β is known (for example, by simulating data). The average, $\langle \bullet \rangle$, is the same as in the previous sections, over repeated measurements.

By the definition of multidimensional Gaussian data

$$P(Y \mid \psi) \propto \exp\left(-\frac{1}{2} \sum_{k,l} H_{k,l} (Y_k - \psi_k)(Y_l - \psi_l) \right) \quad , \tag{17}$$

where $H_{k,l}$ is symmetric and positive definite. This implies, using matrix notation, that the matrix H can be factored as $H = L^T L$, in terms of a non-unique, invertible (perhaps requiring slight augmentation of the diagonal), lower triangle matrix L, where L^T denotes the transpose of L.

By changing variables to $Z = L(Y - \psi)$, we see from Eq. (17) that in terms of Z the data probability has a standard normal form $\propto \exp(-Z^T Z / 2)$, so that $\langle Z_m Z_n \rangle = \delta_{m,n}$. Writing out Eq. (16) in terms of Z, we therefore obtain

$$\left\langle \sum_{m,n} L_{k,m}^{-1} Z_m L_{l,n}^{-1} Z_n \right\rangle = \sum_m L_{k,m}^{-1} L_{l,m}^{-1} = \beta_{k,l} \quad ,$$

or, using matrix notation, $L^{-1}\left(L^{-1}\right)^T = H^{-1} = \beta$, and therefore $H = \beta^{-1}$. The likelihood function corresponding to Eq. (17) is

$$P(Y \mid \psi) \propto \exp\left(-\frac{1}{2} (L(Y - \psi))^T (L(Y - \psi)) \right) \quad . \tag{18}$$

Uncertainty or finite resolution of the forward model calculation.
Finally, we discuss the situation where measurement uncertainty would seem negligible and irrelevant. An example is climate-change modeling of ice sheet melting (MacAyeal et al. 1996), where the measurements of ice shelf thickness and movement velocity are quite precise, but the three-dimensional ice sheet modeling has limited resolution. The uncertainty then is a question of the variability of the measured quantities over the spatial resolution of the model calculation and is like the normalizing coefficient variability in what we have discussed here when the numbers of counts are large. The logarithmic standard deviation of the observed data over a spatial region representing the resolution of the model calculation could be used to determine the lognormal likelihood function S for each such region. In short, a lognormal likelihood function could be used with an empirically determined standard deviation S.

Exercises

1. Assume many independent measurements of the same quantity. Prove that the combined likelihood function, given all the measurements, is the product of the individual likelihood functions.
2. What is the combined likelihood function of n independent measurements, each of which has a normal distribution?
3. For given background counting rate the log of the likelihood function in Eq. (3) is given by $Log(L(\psi)) \propto N \log(A\psi/T + \lambda_B) - (A\psi + \lambda_B T)$. Show that this is not changed if the N time intervals between single counts are measured rather than measuring the number of counts in time interval T? Is there any advantage to taking data in this way?
4. In Chapter 4 the parameters of the distribution of true background counting rate were determined in two ways: 1) by minimizing the sum of the squares of the differences between the empirical and theoretical cumulative probabilities and 2) by maximizing the combined likelihood function. Compare these approaches.
5. Derive the normal approximation for the "less than" likelihood function shown in Fig. 3.
6. Recalculate Figs. 4 and 5 using the Fortran program `SetUpBioData` included with the supplementary material for this chapter.
7. Derive a general expression for the combined likelihood function in terms of $\chi(\psi, Y^{(i)})$ for an individual measurement.
8. Fill in the steps in the derivation of the Empirical Likelihood function given by Eq. (15).

References

1. MacAyeal, D.R., V. Rommelaere, P. Huybrechts, C.L. Hulbe, J. Determann, and C. Ritz. "An Ice-Shelf Model Test Based on the Ross Ice Shelf, Antarctica." *Annals of Glaciology* 23:46–51 (1996).
2. Miller, G. "A Simple and Conservative Empirical Likelihood Function--Corrected." *Electronic Journal of Applied Statistical Analysis* (2015).

Chapter 8. Prior Probability Distributions

In the type of probabilistic analysis of data described in this book, prior probability distributions are always required. The prior represents the experiment that produced the thing that is being measured. It gives the starting distribution of parameter values, before being filtered by the likelihood functions of the measurements. As such, the prior influences the final interpretation of the measurement. What prior should be used?

Location parameter. As a starting point, consider the measurement of the location of a point in space. If the experiment that produces the point makes no reference to any other points in space (translation invariance), the prior must be uniform. This follows from coordinate transformation invariance. It cannot matter what coordinate system is used for the prior. Imagine a one-dimensional system and a new, shifted coordinate system with new coordinate

$$x' = x - x_0 \quad ,$$

(1)

where x_0 is the amount of the shift. In the new coordinate system the prior is denoted by $P'(x')$ and, because the same physical probability can be expressed in either coordinate system,

$$P'(x')dx' = P(x)dx \quad .$$

This implies

$$P'(x') = P(x' + x_0)$$

(2)

for any shift amount x_0. Equation (2) requires that $P(x)$ be a constant, or uniform function.

If the situation under consideration is invariant under a change of scale (everything multiplied by the same scale factor), the prior in terms of the log of the scale factor must be uniform. This follows from

$$x' = x / f$$
$$\log(x') = \log(x) - \log(f)$$

and $\log(f)$ acts like a coordinate system shift in terms of logs.

A prior probability distribution ultimately is justified by experimental data, and these invariance principles refer to the measurement data.

These two invariance principles can reduced to two simple rules of thumb. If the parameter can be both positive and negative, in lieu of other information, the prior is chosen as uniform. If the parameter is always positive, in lieu of other information, the prior is chosen as uniform in terms of the log of the parameter. If the prior needs to impart some information, a broad normal or lognormal is often used in these two situations respectively. The limit of a very broad normal becomes the uniform distribution, and the limit of a very broad lognormal becomes the log-space uniform distribution.

Prior for multiple independent parameters. With multiple parameters, the priors are for simplicity usually chosen to be the product of priors for the individual variables; that is, the prior parameters are assumed to be independent.

The priors for individual parameters can always be transformed to be uniform distributions from 0 to 1 by using θ variables representing the cumulative probability for the parameter. For a uniform prior between ξ_{min} and ξ_{max}, the transformation is

$$\xi = \xi_{min} + \theta(\xi_{max} - \xi_{min}) \quad .$$

For a log-space uniform prior, the transformation is

$$\xi = \xi_{min} \exp\left(\theta \log(\xi_{max} / \xi_{min})\right) \quad .$$

For the alpha prior to be discussed subsequently, the transformation is

$$\xi = \xi_{max} \theta^{1/(\alpha \Delta t)} \quad .$$

For a normal or lognormal, it is

$$\xi = \xi_0 + \sigma \Theta_{norm}^{-1}(\theta)$$
$$\xi = \hat{\xi} \exp\left(\sigma \Theta_{norm}^{-1}(\theta)\right) \quad ,$$

where Θ_{norm}^{-1} is the inverse of the standard normal cumulative probability, a function readily available in subroutine libraries (often named "normsinv").

If other independent measurements exist, the prior is the posterior based on these other measurements.

In cases where there are many measurements or a few measurements with small uncertainty, the influence of the prior is minimal. However, a more extreme and informative prior is also sometimes warranted. Let us discuss this important situation in more detail.

Simple example—white and black atoms. Imagine measurements to ascertain whether atoms are white or black. For this purpose, one uses a digital measuring instrument that reads 0 for a white atom and 1 for a black atom. However, this measurement is not perfect; if a large number of white atoms or black atoms are run, the results shown in Figures 1 and 2 are obtained.

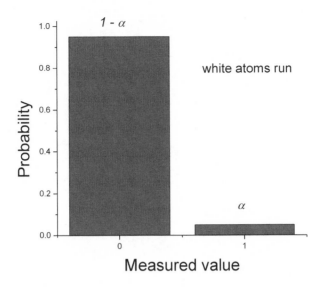

Figure 1—Measurement results for white atoms.

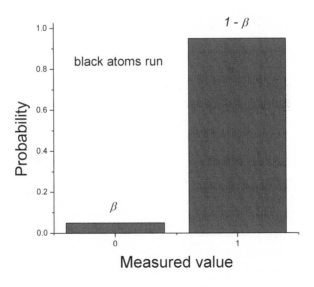

Figure 2—Measurement results for black atoms.

The quantities α and β are the false positive and false negative rates. If these quantities are 0 the measurement is perfect and there is no need for a probabilistic analysis. If these quantities are small, the measuring system is good—but how good? The further analysis required is not difficult.

Let us consider measuring N atoms and enumerate the possible outcomes. Drawing one atom, it is white with probability p and black with probability $q = 1 - p$. The different possibilities are shown in Table 1.

Table 1—*Numbers of different outcomes when* N *atoms are measured.*

color	meas0	meas1
white	$pN(1-\alpha)$	$pN\alpha$
black	$qN\beta$	$qN(1-\beta)$

Now, imagine that we have measured 1. What can we conclude? The fraction of the time that the atom is black, which is the same as the probability, is

$$P(black \mid meas1) = \frac{qN(1-\beta)}{qN(1-\beta)+pN\alpha} \; . \tag{3}$$

This quantity subtracted from 1 is the probability that the atom is white (the atom must be either black or white):

$$P(white \mid meas1) = \frac{pN\alpha}{qN(1-\beta)+pN\alpha} \; . \tag{4}$$

In the above equations, the vertical bar is read as "given" and expresses a conditional probability.

The probability distribution conditioned on the measurement result given by Equations 3 and 4 (called the posterior distribution) is what is almost always desired, rather than simply stating that the measurement indicated black and that the false positive rate is small. The intent of the measurement is to have an interpretation or model in terms of atom color. This is the bottom-line quantity of interest. It is completely natural to want to know this bottom-line result after the measurement is done. In fact, studies have shown that civil-life users of information about measurements usually assume that they are being told about the posterior distribution, whether or not this is the case.

Before any measurements, the probability that an atom is black (that an atom taken from the atom container will be black) is called the prior probability. This distribution is given by

$$P(black) = \frac{N_{black}}{N_{black} + N_{white}} = q$$

$$P(white) = \frac{N_{white}}{N_{black} + N_{white}} = p$$

The probability distribution of atom color is updated after the measurement by using Bayes theorem:

$$P(black \mid meas1) \propto P(meas1 \mid black)P(black)$$
$$P(white \mid meas1) \propto P(meas1 \mid white)P(white)$$

The likelihood function, which is the probability distribution of the measurement results given the atom color, is shown in Figures 1 and 2:

$$P(meas1 \mid black) = 1 - \beta$$
$$P(meas1 \mid white) = \alpha$$

Bayes theorem, of course, reproduces what has been already derived in Eqs. (3) and (4) using elementary reasoning.

Putting in some numbers may be helpful, e.g., for $N = 10000$ atoms,

$$N_{black} = qN = 10$$
$$N_{white} = pN \cong 10000$$
$$\alpha = 0.05$$
$$\beta = 0.05$$

In this case,

$$P(black \mid meas1) = \frac{0.95 \times 10}{0.95 \times 10 + 0.05 \times 10000} = 0.02$$

$$P(white \mid meas1) = \frac{0.05 \times 10000}{0.95 \times 10 + 0.05 \times 10000} = 0.98$$

The posterior distribution clearly summarizes the knowledge of what is desired (atom color). Note that in this case, because black atoms are rare in the measured population, this result completely reverses the naive expectation, based on the smallness of α and β, that if the measurement equals 1, then the atom is likely to be black.

In practice an ambiguous or unexpected result is usually followed up by more measurements. If the measurements are independent and are meas1, meas0,... the posterior probability is given by

$$P(black \mid meas1, meas0,...) \propto P(meas1 \mid black)P(meas0 \mid black)...P(black)$$
$$P(white \mid meas1, meas0,...) \propto P(meas1 \mid white)P(meas0 \mid white)...P(white)$$

. (5)

One can see from Eq. (5) that if enough measurements are taken, eventually the measurements will overwhelm the prior. That is, if one continues to measure black many times, the posterior probability of black will soon become very close to 1.

One also sees from Eq. (5) how preceding measurements provide a prior for measurements that follow.

Alpha prior. Another example (besides the case of rare black atoms) of an extreme, informative prior is provided by the probability of intakes of a radionuclide occurring in a time interval Δt.

We expect the distribution of intakes to depend on the size of the time interval Δt. As Δt increases, there would be more time for intakes to occur and conversely. Furthermore, we expect smaller intakes to be more probable than large intakes.

We consider intake probability functions of the form

$$w(\xi)dt = \frac{\alpha\,dt}{\xi}e^{-\beta\xi} \quad , \tag{6}$$

where α is a probability per unit time, and the parameter β limits the distribution for large values of ξ. This distribution, representing the probability distribution of the sum of intakes ξ in an infinitesimal time interval dt, is improper because the integrated probability diverges for small intakes.

Following the discussion in Chapter 6 of the probability distribution of a particles energy loss as a function of distance of travel though matter, the probability distribution of total intake in a finite time interval Δt is then given by

$$P(\xi, \Delta t) = \frac{\alpha \Delta t}{\xi} (\beta \xi)^{\alpha \Delta t} \frac{\exp(-\beta \xi)}{(\alpha \Delta t)!} \quad , \tag{7}$$

which is a gamma distribution. The function $(x)!$ is the Gamma function, given approximately by

$$x! \equiv \Gamma(1+x) \cong 1 - Cx$$

for $x \ll 1$, where $\Gamma(.)$ is the Gamma function and C is Euler's constant, $C = 0.577$.

We would like to simplify Eq. (7), eliminating the infinite domain of ξ and the exponential. This is the so-called "alpha" prior given by

$$P(\xi, \Delta t) d\xi = \alpha \, \Delta t (\beta \xi)^{\alpha \Delta t} d \log(\xi) \quad . \tag{8}$$

The cumulative probability is given by

$$\Theta(\xi, \Delta t) = \int_0^\xi d\xi' \frac{\alpha \, \Delta t}{\xi'} (\beta \xi')^{\alpha \Delta t} = (\beta \xi)^{\alpha \Delta t} \quad . \tag{9}$$

Normalization of the distribution determines β as $1/\xi_{max}$, where ξ_{max} is the maximum allowed value of ξ, so in this sense ξ_{max} is an unimportant normalization parameter and the single important parameter is α.

The alpha distribution is used as a prior on intake amount in internal dosimetry calculations. The interval Δt represents the time interval when the intakes might have occurred. Not only is this interval sometimes small, but from experience at "clean" facilities it is often known that intakes, while truly happening sometimes, are exceedingly rare. Thus α, the probability of intake per unit time, is very small, perhaps 0.001 per year. The alpha prior is used when there is no information that something unusual happened in the interval Δt (no "incident").

Another possibility, sometimes relevant for very dirty facilities, is when the probability of intakes is not small. This is the situation of chronic intakes, where the probability of intake has a normal distribution about a mean value. Using Eq. (7), it is just a matter of what Δt one considers, and as $\Delta t \to \infty$, one always goes over to a chronic intake situation.

Figure 1 shows a comparison of the alpha distribution from Eq. (8) with the gamma distribution.

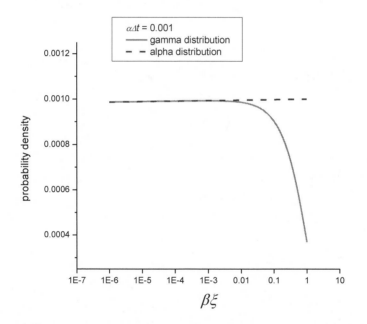

Figure 1—Alpha prior approximation compared with the gamma distribution.

These distributions appear almost constant on a logarithmic scale.

Notice the delta function–like character of Eq. (9). Say $p \equiv \alpha \Delta t = 0.001$ and observe that the probability of intakes up to 0.01 times the maximum is $(0.01)^p = .995$. The probability up to 0.001 times the maximum is $(0.001)^p = .993$. Thus, almost all the probability is concentrated near $\xi = 0$. There is a small tail given by Eq. (8), which is approximately constant in log space,

$$P(\xi, \Delta t)d\xi \cong p\, d(\log(\xi)) \quad . \tag{10}$$

This is what shows up in Fig. 1. The delta function is to the left and off scale.

Consider how this impacts the interpretation of a single bioassay sample. If an intake occurred, the bioassay result Y would be $F\xi$, where ξ is the intake amount and F is some multiplying coefficient giving the expected bioassay result per unit intake from the biokinetic model. The posterior probability of intake amount is given by

$$P(\xi \mid Y) \propto P(Y \mid \xi)P(\xi, \Delta t) \quad ,$$

with $P(\xi, \Delta t)$ from Eq. (9) and

$$P(Y \mid \xi) \propto L(\xi) = \exp\left(-\frac{1}{2}\chi(\xi)^2\right) \quad ,$$

where $L(\xi) = \exp\left(-\chi(\xi)^2 / 2\right)$ is the likelihood function. The posterior distribution therefore has two peaks, one from the prior delta function at zero, and one from the maximum of the likelihood (assuming the likelihood function has a nonzero maximum). The posterior probability of zero intake is therefore

$$P(\xi = 0 \mid Y) = \frac{L(0)}{L(0) + p \int L(\xi) d(\log \xi)} \ .$$

Just as a numerical example, we will assume a normal form for $\chi(\xi)$ (the exact likelihood function should always be used in this situation) so that,

$$\chi(\xi) = \frac{Y - F\xi}{\sigma}$$

$$L(\xi) = \exp\left(-\frac{1}{2}\left(\frac{Y - F\xi}{\sigma}\right)^2\right) \ . \tag{11}$$

If the bioassay result is 2 standard deviations greater than zero $(Y/\sigma = 2)$, one might naively assume that the result is truly positive, meaning that an intake has indeed occurred in the preceding sampling interval. The integral over the maximum of the likelihood function can be estimated as the maximum of the likelihood function from Eq. (11), which is 1, times the factor p from Eq. (10), which is 0.001, times the integration width $\Delta(\log \xi)$, which is approximately $\sqrt{2\pi}\sigma/Y$. The integrated peak at zero is the likelihood function from Eq. (11) evaluated at $\xi = 0$. The posterior probability of zero intake is therefore

$$P(\xi = 0 \mid Y) \cong \frac{\exp\left(-\frac{1}{2}\left(\frac{Y}{\sigma}\right)^2\right)}{\exp\left(-\frac{1}{2}\left(\frac{Y}{\sigma}\right)^2\right) + p\left(\frac{\sqrt{2\pi}\sigma}{Y}\right)} = 0.99 \ .$$

As in the preceding white/black example, our naïve expectation is shown to be incorrect. It is again just a matter taking into account the prior. These kinds of calculations will be done more carefully in Chapter 15.

These considerations are not esoteric or philosophical or even hard to understand. They are just a matter of estimating the prior and properly evaluating the probabilities. Probabilistic modeling naturally does all this.

Exercises

1. Using a spreadsheet, generate 100 values from a normal prior.
2. Generate 100 values from a lognormal prior.
3. Generate 100 values from an alpha prior with $p \equiv \alpha \Delta t = 0.001$.
4. Show that the mean, second moment, and standard deviation of the alpha distribution are given by $\langle \xi \rangle \cong p\xi_{max}$, $\langle \xi^2 \rangle \cong p(\xi_{max})^2/2$,

   $SD \cong p(\xi_{max})^2/2$ with $p \equiv \alpha \Delta t$, which is assumed to be small. Check with the numerical values obtained using the results of the previous exercise. Do they agree? If not why not? How many random numbers from the alpha distribution would have to be summed to produce a normal distribution of the sum like that shown in Fig. 2 of Chapter 5?

Chapter 9. Finite Integer-Valued Markov Chains

With the intent of illustrating the basic concepts in as simple a way as possible, in this chapter we discuss finite integer-valued (discrete) Markov Chains. In this case, straightforward numerical calculations that rely on standard matrix algebra routines can be done.

Markov Chains are sequences of random variables similar to the multiple independent trials considered in Chapter 2, except now the variables are not independent but have a correlation that persists to some degree from trial to trial. In Chapter 2 the alternate mathematical descriptions of probabilities in terms of sequences of random variables and probability distribution functions were discussed. Markov chains generalize these ideas by allowing for correlations between the random variables. Because the sequence number in the chain is now important and is similar to physical time, and also because of the way Markov Chains are used in practice, it is helpful to think of Markov Chains in terms of the average description of some underlying physical system.

Consider an integer-valued function of time $i(t)$ representing the time history of the state (trajectory) of some dynamical physical system. Imagine an ensemble of systems having trajectories corresponding to different initial conditions. The equations of motion, which are unspecified, are assumed to have a stochastic component resulting in complex and irregular trajectories.

Ensemble and time averages. Averaging over a subset of the whole ensemble consisting of trajectories with the same initial condition at $t = 0$, there is some probability $P_i(t)$ that the system is in state i at time t. The ensemble average of some function $g(i(t))$ at time t is therefore given by

$$\langle g(i(t)) \rangle = \sum_i P_i(t) g_i \quad .$$

In practice one calculates time averages of arbitrary functions of the dynamical variable. We assume that the whole ensemble of trajectories for different initial conditions cannot be separated into different groups that have a distinct identity (ergodic assumption). Thus, the long-term time average is the same for all trajectories of the ensemble and independent of initial conditions, so that it can be evaluated as an average over a subset of the ensemble with a particular initialization

$$\lim_{T\to\infty}\left(\frac{1}{T}\sum_{t=1}^{T}g(i(t))\right) = \left\langle\lim_{T\to\infty}\left(\frac{1}{T}\sum_{t=1}^{T}g(i(t))\right)\right\rangle$$

$$= \lim_{T\to\infty}\left(\frac{1}{T}\sum_{t=1}^{T}\langle g(i(t))\rangle\right)$$

$$= \lim_{T\to\infty}\left(\frac{1}{T}\sum_{t=1}^{T}\sum_{i}P_i(t)g_i\right) = \sum_{i}\overline{P}_i g_i$$

where $P_i(t)$ is the ensemble average for a particular initialization, and

$$\overline{P}_i \equiv \lim_{t\to\infty}P_i(t) \quad .$$

Thus, the time average is equal to the ensemble average over the steady-state distribution function. By ergodicity there is a single unique steady-state distribution.

Definition of Markov Chain. The definition of a Markov Chain is that

$$P_i(t+1) = \sum_{j}T_{i,j}P_j(t) \quad , \tag{1}$$

where the conditional transition probabilities $T_{i,j} = P(i;t+1|j;t)$ are independent of time. We use matrix notation to denote the square matrix of transition probabilities T and the column matrix $P(t)$ representing the distribution function at time t.

90

Because probabilities must be normalized,

$$\sum_i P_i(t) = 1$$
$$\sum_i T_{i,j} = 1 \qquad .$$

Also, it is clear from Eq. (1) that

$$P(t) = \underbrace{T \times T \times T ... \times}_{t\ times} P(0) = T^t P(0) \quad , \tag{2}$$

where T^t denotes the t th power of the matrix T (the matrix multiplied by itself t times).

It is assumed that the matrix T can be brought to diagonal form with another square matrix L.

$$T = L\lambda L^{-1} \quad ,$$

where λ is a diagonal matrix of eigenvalues. Therefore

$$T^t = L\lambda^t L^{-1} \quad , \tag{3}$$

which, from Eq. (2), provides a general solution for the distribution function at time t. The columns of L are the eigenvectors $u^{(j)}$ of T satisfying

$$Tu^{(j)} = \lambda_j u^{(j)}$$

so that

$$TL = L\lambda \quad,$$

where λ is a diagonal matrix containing the eigenvalues λ_j along the diagonal.

In practice one constructs a Markov Chain that has as its unique steady state some given distribution function, and one uses chain (time) averages to calculate distribution function averages.

Solution for two-states. Let us consider a specific example.

For two states the most general form of the Markov transition matrix is the following:

$$T = \begin{bmatrix} 1-p_1 & p_2 \\ p_1 & 1-p_2 \end{bmatrix} \quad, \tag{4}$$

where p_1 and p_2 are positive quantities less than or equal to 1. From Eq. (2), $\overline{P}$ is a steady state if and only if

$$T\overline{P} = \overline{P} \quad. \tag{5}$$

Using Eq. (5), p_1 and p_2 are given in terms of the steady-state solution $\overline{P}$, by

$$\frac{p_2}{p_1} = \frac{\overline{P}_1}{\overline{P}_2}$$

92

or,

$$p_2 = a\overline{P}_1$$
$$p_1 = a\overline{P}_2 \quad ,$$

where a is a positive number.

The eigenvalues λ are the roots of the equation

$$\det(T - \lambda I) = (1 - a\overline{P}_2 - \lambda)(1 - a\overline{P}_1 - \lambda) - a^2\overline{P}_1\overline{P}_2$$
$$= (\lambda - 1)(\lambda - 1 + a) = 0 \quad , \tag{6}$$

where I is the identity matrix, expressing the fact that for the equation $(T - \lambda I)u = 0$ to admit a nonzero solution u, the matrix $T - \lambda I$ must be singular. The eigenvalues λ are therefore

$$\lambda = 1, \ 1 - a \ .$$

One can see that the eigenvectors (without a particular normalization) are

$$u^{(1)} = \begin{bmatrix} \overline{P}_1 \\ \overline{P}_2 \end{bmatrix}$$
$$u^{(2)} = \begin{bmatrix} 1 \\ -1 \end{bmatrix} \ .$$

Therefore the matrices L and L^{-1} are given by

$$L = \begin{bmatrix} \overline{P}_1 & 1 \\ \overline{P}_2 & -1 \end{bmatrix}$$

$$L^{-1} = \begin{bmatrix} 1 & 1 \\ \overline{P}_2 & -\overline{P}_1 \end{bmatrix}$$

Let us consider an arbitrary initial distribution given by

$$P(0) = \begin{bmatrix} P_1(0) \\ P_2(0) \end{bmatrix}$$

and ask what the distribution is at time t. The answer is provided by Eqs. (2) and (3) and is given by

$$P(t) = \overline{P} + \lambda_2^t (P(0) - \overline{P}) \quad , \tag{7}$$

where $\lambda_2 = 1 - a$ is the second largest eigenvalue (the largest is always 1). Notice in Eq. (7), because $\lambda_2^t = (\text{sgn}(\lambda_2))^t \exp(t \log(|\lambda_2|))$, where $\text{sgn}(.)$ gives the sign of the argument, the transient power term decays exponentially with time constant

$$\tau = -\frac{1}{\log(|\lambda_2|)} \quad , \tag{8}$$

which is a long time if $|\lambda_2|$ is close to 1, unless the chain starts out at $\overline{P}$. After this transient period, the distribution becomes a unique steady state $\overline{P}$ independent of the starting distribution.

For this example, the transition matrix is given by

$$T = \begin{bmatrix} * & a\overline{P}_1 \\ a\overline{P}_2 & * \end{bmatrix} \quad ,$$

where * denotes what is required by column normalization. In order that the off-diagonal matrix elements remain less than 1 so as to allow column normalization,

$$a \leq \frac{1}{\max(\overline{P}_1, \overline{P}_2)} \quad .$$

The shortest time to reach a steady state occurs when

$$a = \frac{1}{\overline{P}_1 + \overline{P}_2} \quad ,$$

in which case the second eigenvalue is 0. In this case the distribution function becomes the steady state in only one time step, no matter what the starting point. These formulas for a work whatever the normalization of $\overline{P}_1$ and $\overline{P}_2$.

For this example, notice that the chain steady-state distribution satisfies the detailed balance condition

$$T_{1,2}\overline{P}_2 = a\overline{P}_1\overline{P}_2 = T_{2,1}\overline{P}_1 \quad , \tag{9}$$

which, in words means that in steady state, the number of transitions from state 1 to 2 is balanced by the number of transitions from 2 to 1.

In general, the distribution function satisfies the equation

$$\frac{\Delta P_i}{\Delta t} \equiv (TP - P)_i = \sum_j \left(T_{i,j} P_j - T_{j,i} P_i \right) \quad,$$

where $\Delta t = 1$ is the time step, so that detailed balance implies a steady state. One sees immediately that if $TP = P$, P is a steady state, and this is equivalent to detailed balance.

Uniqueness of the steady state. Now we consider the uniqueness of the steady state more carefully.

Imagine that the transition matrix is given by

$$T = \begin{bmatrix} * & a\overline{P}_1 & & \\ a\overline{P}_2 & * & & \\ & & * & b\overline{P}_3 \\ & & b\overline{P}_4 & * \end{bmatrix} . \tag{10}$$

Clearly in this case there is a steady state $\overline{P}$, but there are two eigenvalue-1 eigenvectors, one involving indices 1 and 2 and another involving indices 3 and 4, and the final steady state is not unique. In this case the transition matrix is not ergodic, ergodic meaning that any state can be reached from any other state in a finite number of steps. We see from Eq. (10) that the state with index 3 mixes only with state 4 and not with 1 and 2. In general, if the matrix is ergodic, there is a single unique eigenvalue-1 steady state.

Finding a transition matrix that gives a desired steady state distribution. The actual problem we will be concerned with is to determine the transition matrix T that gives a specified steady state $\overline{P}$. Then, the average of a function over $\overline{P}$ can be evaluated as a chain (time) average of the function of the dynamical variable.

There are many ways to come up with a transition matrix that gives a specified steady state. Such transition matrices are constructed using the detailed balance condition, Eq. (9), together with the requirement that the columns of T contain positive numbers normalized to unity. Detailed balance implies that elements of $T_{i,j}$ reflected through the diagonal (the transposed elements) are in the inverse ratio of the desired steady-state distribution function

$$\frac{T_{i,j}}{T_{j,i}} = \frac{\overline{P}_i}{\overline{P}_j} \quad .$$

In fact, the most basic solution T of $T\overline{P} = \overline{P}$ contains just two nonzero off-diagonal elements in corresponding transposed positions, satisfying the detailed balance relation with neither exceeding unity. The diagonal element in both columns is then the complement of the corresponding off-diagonal element. The other diagonal elements are unity. Other solutions can be constructed as probabilistic averages of these basic transition matrices using the fact that a mixture of solutions is itself a solution (see Exercise 1).

In our consideration of some explicit solutions, we start with a simple case of a random-walk transition matrix T where, from state j, transitions occur only to one other state $i = j+1$ or $i = j-1$. The template of nonzero elements of the transition matrix is therefore the diagonal plus one other adjoining element in each column.

For a basic algorithm to serve as a starting point, within the nonzero template let the j th column of T be given by

$$\begin{aligned} T_{i,j} &= \overline{P}_i \\ T_{j,j} &= * \end{aligned} \quad , \tag{11}$$

where $*$ denotes what is required by column normalization. One sees immediately that the detailed balance condition is satisfied.

The above remains a solution if the off-diagonal matrix elements are multiplied by a positive constant a less than 1. This multiplying factor need be constant only with respect to transposed positions within the matrix T, and it can be greater than 1 as long as it does not upset the column normalization condition.

The diagonal band can be increased from one adjoining state to $l = 2,3,...$ adjoining states, and this form provides a solution when the off-diagonal matrix elements are given by Eq. (11) multiplied by a constant a/l. One should note that for $l = 1$, even though satisfying the detailed balance requirement, the transition matrix is not ergodic, because the transition matrix in that case appears as

$$
\begin{matrix}
* & - & & & \\
- & * & & & \\
& & * & - & \\
& & - & * & \\
& & & & \cdots \\
\end{matrix} \quad .
$$

The solution given by Eq. (11) has the important drawback that it requires the normalized steady-state distribution function. Nonetheless, we will use it to study the effect of the multiplying factor a.

Table 1 shows the result of a numerical study using the Fortran program MC (included in supplementary material for this Chapter). The discrete space has 100 points and the assumed steady-state distribution function is a Gaussian centered on $i = 50$ with a standard deviation of 20. The transition matrix and the 100 eigenvalues and eigenvectors are calculated numerically for a symmetrical random walk where the half width of the random-walk step is 1 ($l = 2$) with different values of the constant a. The eigenvalues are real and positive having a maximum value of 1 for one eigenvector. That eigenvalue-1 eigenvector, when normalized, returns the assumed steady state. The second largest eigenvalue gives the time to reach steady state from Eq. (8). Table 1 shows the times 3τ for three different values of a.

a	3τ
10	1.1×10^4
1	1.1×10^5
0.1	1.1×10^6

It seems that the multiplying factor a should be as large as possible without upsetting column normalization in order to have the smallest times to reach steady state.

Metropolis-Rosenbluth-Teller and Barker algorithms. We can let the quantity a be specific to a particular transposed pair of off-diagonal matrix elements of T rather than being a constant, and a solution so modified will remain a solution. For the Barker (B) algorithm

$$a_{i,j} = a_{j,i} = \frac{1}{l}\frac{1}{\overline{P}_i + \overline{P}_j} \quad , \tag{12}$$

while for the Metropolis-Rosenbluth-Teller algorithm (MRT) it is given by

$$a_{i,j} = a_{j,i} = \frac{1}{l}\frac{1}{\max(\overline{P}_i,\overline{P}_j)} \quad . \tag{13}$$

Notice that the MRT factor from Eq. (13) is always greater than the B factor from Eq. (12). This also corresponds to the relative size of the times to reach steady state given in Table 2.

algorithm	3τ
B	4.4×10^3
MRT	2.3×10^3

The MRT algorithm can be viewed as the maximally efficient member of the family of exact solutions of detailed balance of this type, as has been shown by Peskun (Peskun 1973). It is important to notice that these algorithms do not require the normalized distribution function.

Multiple-candidate algorithm. Another "multiple candidate" (MC) algorithm (Miller 2011) seeks rapid convergence and compatibility with parallel processing, sometimes in exchange for exact detailed balance. For the MC algorithm the columns of the transition matrix are given, within the nonzero template, by

$$T_{i,j} \propto \overline{P}_i \quad , \tag{14}$$

including the diagonal element, with the proportionality constant chosen to provide column-by-column normalization. This solution provides a proper Markov Chain transition matrix, but the steady state now may be only approximately given by the desired distribution function. When the size of the diagonal band in the nonzero template (the size of the random-walk step) is many times larger than the size of the support region of the desired steady-state distribution function, we would expect the approximation to be quite good, and when the diagonal band encompasses the entire space, detailed balance is satisfied exactly. In that case the MC algorithm can be viewed as a numerical implementation of Gibbs sampling (Geman and Geman, 1984; see Exercise 8).

This is illustrated by the following numerical example shown in Fig. 1. The discrete space has 100 points and the assumed distribution function is a Gaussian centered on $i = 50$ with a standard deviation of 4. Two cases are considered, symmetrical random walks where the half width of the random-walk step is 1 ($l = 2$) and 10 ($l = 20$). The transition matrix and the 100 eigenvalues and eigenvectors for the three algorithms are calculated numerically. The eigenvalue-1 eigenvector gives the steady state, and this eigenvector is compared with the desired steady-state distribution function. For the MRT and B algorithms the agreement is exact, as it must be. For the MC algorithm, when the random-walk step size is small, the actual steady state is significantly narrower than the assumed distribution. On the other hand, when the random-walk step size is larger, the agreement is quite good.

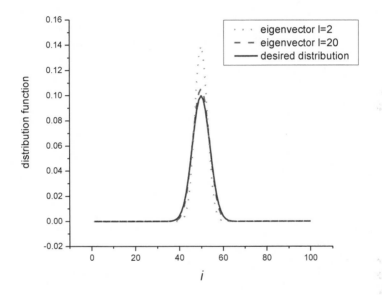

Figure 1—Steady-state distribution functions obtained using the MC algorithm.

It is of interest to note that for $l = 1$, the MC algorithm is exactly the same as the B algorithm and is an exact solution of the detailed balance condition, even though for $l = 1$ the transition matrix is not ergodic except when the entire space has only two states, as in the simple example already discussed.

When investigating these algorithms numerically, one finds that for a wide diagonal band, the rate of convergence for the MC algorithm is much better than for the other two algorithms. This is illustrated in Fig. 2, which shows the eigenvalue spectrum for the three algorithms. Although not clearly visible in the figure, the eigenvalue for $i = 1$ is always 1.

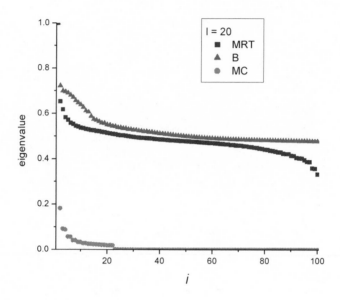

Figure 2—Eigenvalue spectrum for the three different algorithms. The MC algorithm converges much more rapidly than the others.

The transient states decay like λ^t, where t is the chain iteration number and λ is the eigenvalue. The initial state is some mixture of the 100 eigenvectors. So, we desire small values of the non-unity eigenvalues in order for the system to quickly reach a steady state.

Candidate distribution. In practice for Markov Chain Monte Carlo with continuous variables, the transition probabilities are not used directly, but transitions are generated using a two-step process by first generating a "candidate" for the new point from a candidate distribution and then probabilistically accepting or rejecting this new point. If the candidate is rejected, the chain stays fixed at the original point.

Assume that the system is in state j. Let us assume that the new state i is probabilistically generated by first generating a candidate point i with probability $q_{i,j}$ in the diagonal band excluding the center point and then accepting this new state with probability $\alpha_{i,j}$. If the new state is not accepted, the system remains in state j. The probability of the two-step process leading to a change of state (candidate and acceptance) is the product of the two probabilities

$$T_{i,j} = q_{i,j}\alpha_{i,j} \quad . \tag{15}$$

Equation (15) applies for off-diagonal matrix elements. If no candidate is accepted, the system remains in state j. The probability of no change is given by the compliment of the probabilities given by Eq. (15) (that is, by column normalization). Notice that the columns of q are probability distributions over the diagonal band of width l excluding the actual diagonal, while the columns of T are probability distributions over all $l+1$ nonzero elements in the diagonal band of the nonzero template.

So far we have assumed a uniform random-walk candidate distribution with

$$q_{i,j} = \frac{1}{l} \quad . $$

The acceptance probability can be obtained from the transition matrix using Eq. (15). From Eq. (13) the acceptance probability for the MRT algorithm in this case can be written as

$$\alpha_{i,j} = \min\left(1, \frac{\overline{P}_i q_{j,i}}{\overline{P}_j q_{i,j}}\right) \quad . \tag{16}$$

Equation (16) provides a more general form of the MRT algorithm (Hastings 1970).

Notice that the acceptance probability is less than or equal to 1, which means that column normalization of T goes through, because the column-by-column summation of off-diagonal terms of T will be less than 1.

We can now reconsider the MC algorithm. Let us assume a candidate distribution given by

$$q_{i,j} = \frac{\overline{P}_i}{\sum_{j' \approx j} \overline{P}_{j'}} \quad , \tag{17}$$

where $j' \approx j$ indicates that the summation is over l indices around, but excluding j. Equation (17) would seem to be a candidate distribution that makes optimal use of the information provided by having multiple candidates, because candidates are concentrated where $\overline{P}$ is large. From Eqs. (15) and (16), using this candidate distribution,

$$T_{i,j} = \frac{\overline{P}_i}{\sum_{j' \approx j} \overline{P}_{j'}} \min\left(1, \frac{\sum_{j' \approx j} \overline{P}_{j'}}{\sum_{j' \approx i} \overline{P}_{j'}}\right) \quad , \tag{18}$$

instead of

$$T_{i,j} = \frac{\overline{P}_i}{\overline{P}_j + \sum_{j' \approx j} \overline{P}_{j'}} \quad , \tag{19}$$

for the MC algorithm. Equation (18) is approximately the same as Eq. (19) in the limit where the diagonal band is large, in which case both summations in Eq. (18) are approximately the same and much larger than the diagonal term included with the summation in the denominator of Eq. (19).

The algorithm given by Eq. (18) corrects the detailed balance deficiency of the MC algorithm. This algorithm is a uniform-proposal-distribution version of the MTM(II) algorithm discussed by Liu et. al (2000). We find it to be not a useful algorithm unless the diagonal band encompasses the entire space, because its convergence can be very slow. For a random walk candidate distribution, the ratio in Eq. (16) is $\overline{P}_i / \overline{P}_j$. Thus, when the current chain position j is far out in the tail of the steady state distribution, the acceptance probability will be 1 in moving toward the peak. However in Eq. (18) the ratio is reversed in this sense, and the numerator involves points near j, while the denominator involves points near i. As shown in Fig. 3, when the point j is far outside the support region of the steady-state distribution function, the summation with $j' \approx j$ is much smaller than the summation with $j' \approx i$ and the chain has very low probability of transitioning from $j \rightarrow i$.

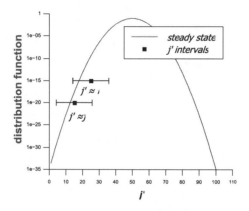

Figure 3—Illustration of the slow convergence of the Eq. (18) MRTH algorithm. When the point j is far outside the support region of the steady-state distribution function, the sum $j' \approx j$ is much smaller than the sum $j' \approx i$, and the chain has very low probability of transitioning from $j \rightarrow i$.

As before, $l = 20$ and the assumed steady-state distribution is a Gaussian centered at $i = 50$ with standard deviation 4. When the diagonal band encompasses the entire space, both algorithms converge very quickly, but the Eq. (19) MC algorithm is optimum, because the denominator is 1, and the algorithm is generating independent samples from the posterior distribution.

Figure 4 shows the distribution after 7 iterations with an initial state $j = 15$ for the MC algorithm, the MRT algorithm, and the MRTH algorithm defined by Eq. (18).

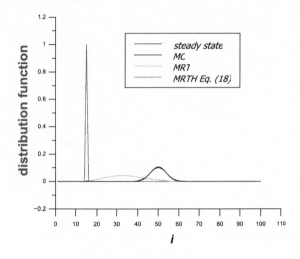

Figure 4—Distribution, with the chain initially at $j = 15$, after 7 steps.

With the initial state concentrated at $j = 15$, the MC algorithm moves the distribution toward the steady state as much as possible given the size of the diagonal band, while the Eq. (18) MRTH algorithm leaves it almost unchanged.

Parallel processing. The simulation of the dynamics consists in first generating a candidate point, calculating the acceptance probability for this candidate, and, if the candidate is accepted, moving to the new point. Because the transition probability just involves the desired distribution function evaluated at the points j and i, this process requires one new evaluation of the desired distribution function at the candidate point i.

From the standpoint of parallel processing, we would like to be able to take advantage of more than one evaluation of the desired distribution function for each iteration of the chain, because these calculations, which

106

usually involve the bulk of the computer time, could then be performed in parallel. To do this, the l transition probabilities $T_{i,j}$ from point j to all the possible new off-diagonal points i can be calculated in parallel (in the continuous variable case, a new distribution function evaluation is needed for each off-diagonal point), and the new chain position is then probabilistically generated from this discrete probability distribution. This parallelization method works for any of the algorithms, but it is demonstrated in Chapter 11 that parallelization is only advantageous for the MC algorithm. This is also illustrated to some extent in Table 3 below, which shows the times required to reach a steady state from Eq. (8) for the sample problem we have been considering (assumed steady-state distribution is a Gaussian centered at $i = 50$ with standard deviation 4). Increasing the size of the diagonal band (l) improves convergence times for any of the algorithms, but beyond that the MC algorithm is significantly better, in sharp contrast with the detailed balance MC algorithm given by Eq. (18).

Table 3—Number of chain iterations required to reach steady state (3τ) with different numbers of multiple candidates (l) for four algorithms.

algorithm	l	3τ
MRT	2	106
B		193.5
MRT	20	7.1
B		9.2
MC		1.8
MRTH Eq. (18)		5×10^9

Exercises

1. Assume two different transition matrices $T^{(1)}$ and $T^{(2)}$ that have a desired steady state $\overline{P}$ but are not ergodic. Compare the mixed transition matrix $(T^{(1)} + T^{(2)})/2$ with the compound transition matrix $T^{(1)}T^{(2)}$ as possible ergodic transition matrices.

2. Using a spreadsheet, construct an integer-valued dynamical system that has a specified steady-state distribution function where the dynamical variable can assume only the two values 0 or 1. Hint: consider using the immediate if function IF(LOGICAL TEST, VALUE IF TRUE, VALUE IF FALSE).

3. Using a time (chain) average, find the average value of the integer variable in the previous exercise. Compare with the value of the desired distribution function $\overline{P_1}$. How does this average value depend on the chain starting point or different random number sequences?

4. Now multiply $\overline{P_1}$ and $\overline{P_2}$ by a positive factor a. How does the chain average and the standard deviation of the chain average depend on the positive multiplying factor a?

5. Derive a formula for the standard deviation in Exercise 4 as a function of a. Compare with the data from Exercise 4.

6. What happens to the Markov Chain constructed in problem 2 when $a = 1$? Show that the sequence in this case consists of independent trials. How large can the multiplying factor a be? What happens when $\overline{P_1} = \overline{P_2} = 0.5$ and $a = 2$?

7. Recalculate the numbers in Table 1 using the Fortran program MC included in the supplementary material for this chapter.

8. Show how the MC algorithm can be viewed as Gibbs sampling.

References

1. Barker, A. A. "Monte Carlo Calculations of the Radial Distribution Functions for a Proton-Electron Plasma." *Australian Journal of Physics* 18:119–133 (1965).

2. Miller, G. "Markov Chain Monte Carlo Algorithms Allowing Parallel Processing –II." *The Open Numerical Methods Journal* 3:12–19 (2011).

3. Peskun, P. H. "Optimum Monte-Carlo Sampling Using Markov Chains." *Biometrika* 60(3):607–612 (1973).

4. Geman, S. and D. Geman. "Stochastic Relaxation, Gibbs Distributions, and the Bayesian Restoration of Images." *IEEE Trans Patt Analysis Mach. Intel* PamI-6(6):721-741 (1984).

5. Hastings, W. K. "Monte Carlo Sampling Methods Using Markov Chains and their Applications." *Biometrika* 57(1):97–109 (1970).

6. Liu, J. S., Liang, F. and W. H. Wong, "The Multiple-Try Method and Local Optimization in Metropolis Sampling." *Journal of the American Statistical Association,* **95**(449):121–134 (2000).

Chapter 10. Markov Chain Monte Carlo for Continuous Variables

In this chapter understanding of discrete Markov Chains will be applied to the actual problem, which involves continuous variables.

There is a possibly multidimensional parameter space with parameter point denoted by θ. Each of the dimensions j of θ has domain 0 to 1. The problem of interest may be stated in terms of multidimensional integration that is used to evaluate average values of an arbitrary function with respect to a given probability distribution. The goal is to calculate distribution function averages of the form

$$Avg(F) \equiv \langle F \rangle = \frac{\int d\theta\, P(\theta) F(\theta)}{\int d\theta\, P(\theta)} \quad , \tag{1}$$

where F is an arbitrary function of θ, the probability distribution $P(\theta) \propto e^{-E(\theta)}$, and the given "energy" function $E(\theta)$ is bounded from below. In our case

$$E(\theta) = \frac{\chi^2(\theta)}{2} \quad .$$

The basic method used to solve this problem (Metropolis, Rosenbluth, and Teller 1952) is Markov Chain Monte Carlo integration, where an algorithm produces a Markov Chain with a stationary state $P(\theta)$. The distribution function average given by Eq. (1) is then approximated by the chain average

$$\langle F \rangle \cong \frac{1}{T} \sum_{t=0}^{T} F(\theta_t) \quad , \tag{2}$$

for large $T \to \infty$, where t enumerates the chain iteration.

Candidate distribution assuming multiple-candidates. The Markov Chain algorithm (the rule telling the computer how to select the next point θ' in parameter space, given that the chain is at a current point θ) is the following: First l candidates for the new point, labeled by

$i = 1, \ldots l$, are generated from a conditional probability distribution $q(\theta' \mid \theta)$ (read as "the probability distribution of θ' given, or conditioned on, θ"). This "candidate" probability distribution is chosen to be a random walk, $\theta' = \theta + \Delta(2x - 1)$, where x is a random number uniformly distributed between 0 and 1, and Δ is a fixed parameter, for all or some subset of the dimensions of θ'. The "energy" calculations for the l candidates can be done using parallel processing.

The candidate is accepted and the chain is moved to the new point $i = 1, \ldots l$ with probability $\alpha(\theta'_i, \theta)$. Only a single one of the l candidates will be probabilistically accepted as the next position of the chain. If no candidate is accepted, the chain remains at its current position. The probability of no chain movement is given by

$$\alpha(\theta, \theta) = 1 - \sum_{i=1}^{l} \alpha(\theta'_i, \theta) \quad .$$

Three multiple-candidate Markov-Chain algorithms. Three different Markov Chain algorithms are considered. These will be stated in terms of the candidate distribution and the desired steady-state distribution function P.

1) For the Barker (B) algorithm

$$\alpha(\theta'_i, \theta) = \frac{1}{l} \frac{\dfrac{P(\theta'_i)}{q(\theta'_i \mid \theta)}}{\dfrac{P(\theta'_i)}{q(\theta'_i \mid \theta)} + \dfrac{P(\theta)}{q(\theta \mid \theta'_i)}} \quad . \tag{3}$$

One can verify the detailed balance condition

$$\frac{q(\theta \mid \theta'_i)\,\alpha(\theta \mid \theta'_i)P(\theta'_i)}{q(\theta'_i \mid \theta)\,\alpha(\theta'_i \mid \theta)P(\theta)} = 1 \quad .$$

2) The Metropolis-Rosenbluth-Teller algorithm, as generalized to include a general candidate distribution by Hastings[4] (MRTH), has

$$\alpha(\theta_i', \theta) = \frac{1}{l} \min\left(\frac{P(\theta_i')}{q(\theta_i' \mid \theta)} \frac{q(\theta \mid \theta_i')}{P(\theta)}, 1 \right) \quad , \tag{4}$$

and one can similarly verify detailed balance.

3) For the Multiple Candidate (MC) algorithm,

$$\alpha(\theta_i', \theta) = \frac{\dfrac{P(\theta_i')}{q(\theta_i' \mid \theta)}}{\displaystyle\sum_{k=0}^{l} \frac{P(\theta_k')}{q(\theta_k' \mid \theta)}} \quad . \tag{5}$$

For $k = 0$ in the above, θ_k' is to be replaced by θ. The detailed balance relationship is

$$\frac{q(\theta \mid \theta_i')\alpha(\theta \mid \theta_i')P(\theta_i')}{q(\theta_i' \mid \theta)\alpha(\theta_i' \mid \theta)P(\theta)} = \frac{\displaystyle\sum_{k=0}^{l} \frac{P(\theta_k')}{q(\theta_k' \mid \theta)}}{\displaystyle\sum_{k'=0}^{l} \frac{P(\theta_{k'}')}{q(\theta_{k'}' \mid \theta_i')}} \quad . \tag{6}$$

In the random-walk situation q is simply either 0 or a constant

$$q(\theta' \mid \theta) = \frac{1}{2\Delta} \quad ,$$

where $|\theta' - \theta| < \Delta$ for each dimension of θ, and it drops out of the formulas. For a random-walk candidate distribution, the points θ_k' for the sum in the numerator of Eq. (6) are situated within Δ of the central point θ and the points $\theta_{k'}'$ for the sum in the denominator are situated within Δ of one of these points θ_i'. The ratio is approximately equal to 1 for any l when the region explored by the candidate distribution is large enough to encompass the support region of the desired steady-state distribution function P.

A single Markov Chain is run; however, for each iteration l candidates for the next position of the chain are generated. Only a single one of these candidates will be probabilistically accepted as the next position of the chain. If no candidate is accepted, the chain remains at its current position.

Note that for the MRTH and B algorithms with multiple candidates, the acceptance probability is just the average over the candidates of the original, single-candidate expressions.

The algorithms stated here use l candidates at each iteration of the chain, the energies of which can be computed simultaneously using parallel processing. The MRTH and B algorithms, as originally stated, generated only a single new candidate. The multiple candidate versions of the MRTH and B algorithms are given just for academic interest, because having multiple candidates does not improve the performance of these algorithms, as will be discussed in the next chapter.

Using a random walk candidate distribution, the MRTH algorithm is the same as the original MRT algorithm, so we will refer to it in this way. Also, generally the MRT algorithm is more efficient than the B algorithm, so in practice we will use either the MRT algorithm with $l = 1$, or the MC-l algorithm with l multiple candidates.

Exercises

1. Derive Eq. (6).
2. Show how to use this continuous formalism for discrete variables.

Reference

1. Hastings, W. K. "Monte Carlo Sampling Methods Using Markov Chains and Their Applications." *Biometrika* 57(1):97–109 (1970).

Chapter 11. Markov Chain Monte Carlo— Practical Aspects

This chapter discusses some practical aspects of MCMC calculations. The Fortran program that does the MCMC calculations discussed in this chapter is part of the supplementary material for this chapter.

Convergence. One needs to know how long to run the chain for convergence to the steady state. This can be assessed in practice by running two independent chains with different starting points and different random number seeds and comparing the results. If more than two chains are run, all the chains run are divided into two equal-length combined chains. The comparison of the desired distribution (of some arbitrary $F(\theta)$) calculated from the two combined chains directly gives the degree of convergence. To minimize the effects of initial transient states ("initialization bias"), some initial fraction (the "burn-in" fraction) of each chain is discarded. Convergence is obtained by increasing the total number of iterations, with the same fraction of initial iterations discarded, until the two independent chains give an acceptable agreement for the distributions of the quantities of interest. The agreement is assessed by comparing the standard deviation of the averages of the quantities from the two runs (for only two samples x_1 and x_2 the standard deviation is given by $(x_1 - x_2)/\sqrt{2}$) with the average of the two standard deviations, which gives the uncertainty of the quantity intrinsic to the problem. The idea is that the difference between the two runs should be small compared with the intrinsic uncertainty of the quantity. In practice we require that the standard deviation be less than $1/3$ of the average over the two runs for the quantities of interest,

$$\frac{|\,Avg(F(\theta))_1 - Avg(F(\theta))_2\,|}{\sqrt{2}} < \frac{1}{3} \frac{SD(F(\theta))_1 + SD(F(\theta))_2}{2} \,.$$

Tape file. The "result" of the MCMC run is nicely captured as a "tape" file consisting of model parameter values uniformly spaced throughout the run. The number of iterations recorded in this file is fixed and determined by storage requirements (if many cases are to be calculated and stored) and the final accuracy desired. Usually 100 to 1000 records are sufficient. While the total length of the run is being increased until convergence is obtained, each record of the tape file represents a larger and larger number of chain iterations, and the model parameters in the tape file will approach an independent sample from the posterior distribution.

When the forward model is time consuming, for example when it involves solving a system of differential equations in time, interpolation tables for forward model calculations using the model parameters in the tape file can be saved, which can be viewed as an extension of the tape file, such that forward model calculations never need to be repeated.

One can look at distributions of any desired quantities dependent on the model parameters by reading the tape file, calculating the quantities of interest, and displaying the distributions, without having to rerun the chain—a quick process.

The tape file is the "interpretation of the data in terms of the model." It is a collection of some large number (say 100 to 1000) of alternate possible interpretations of the data in terms of model parameter values.

The tape file needs to include information from the random number generator to allow a continuation of the same sequence of random numbers if it is desired to continue the run. A continued run will give exactly the same answer as a single long run.

To continue a run because of lack of convergence, the new run should be a significant factor longer—for example, a factor of 2—at each stage of nonconvergence. Note that some quantities (those that are effectively averages) will converge earlier than others. The algorithm used in the later chapters of this book runs two chains starting with some nominal chain length found by experience to be about the minimum required, checks all the quantities of interest for convergence and, if any quantity is non-convergent, increases the number of chain iterations by a factor of 2 and repeats the process until convergence is obtained.

One-dimensional simple example. We now discuss a simple example, thinking of it as a problem in numerical integration. The Fortran program used, called MCMC, is included in the supplementary material for this chapter.

In the original Metropolis-Rosenbluth-Teller paper, the problem in statistical mechanics that was being addressed required the evaluation of an integral over a high-dimensional configuration space, representing the positions of N particles, of the function $\exp(-E/kT)$, where E is the potential energy of the system as a function of the $2N$ particle coordinates (a two-dimensional space was assumed), the temperature T is a constant, and k is Boltzmann's constant. The authors stated:

It is evidently impractical to carry out a hundred-dimension integral by the usual numerical methods, so we resort to the Monte Carlo method. The Monte Carlo method for many-dimensional integrals consists simply in integrating over a random array of points instead of over a regular array of points. Thus the most naïve method of carrying out the integration would be to put each of the N particles at a random position in the square (this defines a random position in the $2N$ -dimensional configuration space), then calculate the energy of the system..., and give this configuration a weight of $\exp(-E/kT)$. *This method, however, is not practical..., since with high probability we choose a configuration where* $\exp(-E/kT)$ *is very small; hence a configuration of very low weight. So the method we employ is actually a modified Monte Carlo scheme, where instead of choosing configuration randomly, and weighting them with* $\exp(-E/kT)$, *we choose configurations with a probability* $\exp(-E/kT)$ *and weight them evenly.*

The "most naïve" method of Monte Carlo integration works well in many situations and was used to calculate the exact likelihood function in Chapter 7. However, consider this situation: The energy function or likelihood is a simple Gaussian, however one that is very narrow. In terms of the θ parameter (that runs from 0 to 1), the width of the Gaussian is 10^{-6}. Even though this problem is only one-dimensional and very simple, it simulates the basic problem of exploring a relatively vast parameter space (the region of interest is a minute fraction of the entire space). A one-dimensional integration algorithm, even an adaptive grid algorithm, which adds extra grid points where needed, needs a very fine starting mesh in order to succeed in this situation. Naïve Monte Carlo integration would require about 10^8 random points to have 100 of them land in the vicinity of the likelihood function. Using the MRT algorithm, the results for the energy history shown in Fig. 1 are obtained.

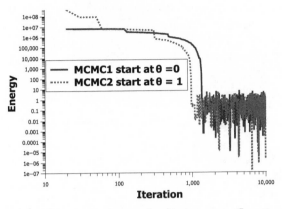

Figure 1—Energy versus iteration number for the narrow Gaussian example.

After about 1000 iterations the chain has "equilibrated" and is fluctuating around the minimum energy point, having moved from the starting point. For the calculation shown in Fig. 1, $\Delta = 1$ half the time and the other half $\Delta = 15 \times 10^{-6}$ or 15 times the standard deviation σ of the Gaussian.

Starting from a point far from the minimum, if Δ is large the chain makes large steps toward the minimum. However, when near the minimum, large Δ leads to most candidates having much higher energy and not being accepted, and the chain then seldom moves. Near the minimum a small Δ is needed for the desired higher acceptance fraction. Figure 2 shows the energy history for different Δ's.

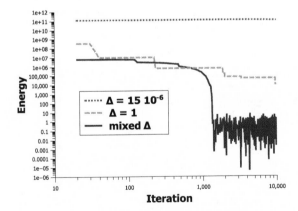

Figure 2—Effect of the random-walk half-step size Δ for the narrow Gaussian example.

If the random-walk half-step size is large, $\Delta = 1$, the fraction of iterations where the chain moves is small (acceptance fraction small), and the chain will stay fixed for many iterations before moving. On the other hand, if Δ is small, the chain takes only tiny steps away from the initial location. Mixed Δ means half the time Δ is large and half small. This technique is somewhat inefficient in that, after "equilibration," for those times when the half-step size is large, the chain almost never moves, so these iterations are wasted.

One might choose to have an initial "adaptive phase" of the algorithm where Δ is sequentially decreased until a desired value of the acceptance fraction is obtained. However, there are pitfalls around decreasing Δ to have a desired acceptance fraction, as will be discussed. Or, one might use a nonlinear least-squares algorithm (discussed in the next chapter) to get into the modal region before starting the MCMC calculation. Or, one might just accept the factor of 2 inefficiency of the mixed Δ method, noting that there is safety, in terms of exploring the entire space, in having Δ large.

In this book, the acceptance fraction is not a specified quantity but a secondary quantity that enters in through the requirement that the number of chain iterations times the acceptance fraction, that is, the number of chain moves, should be large enough (say 100 to 1000), the same or larger than the number on the tape file, in order to have parameter variation in the tape file. If this requirement is not met, the number of chain iterations needs to be increased.

Six-dimensional example. Now we modify the example by adding dimensions. We will use 6 dimensions and the likelihood function width (SD) in each dimension is the same 10^{-6}. Nested one-dimensional integration is out of the question, and naïve Monte Carlo would require 10^{48} random points.

An important technique with MCMC is to separate the parameters into groups and move the parameters one group at a time rather than all together. At each iteration, the chain focuses on only one group. This parameter group can be chosen cyclically or probabilistically, using some time-invariant rule. For the candidates, only this one parameter group is moved. The other parameters are left at their old values.

Figure 3 shows the effect of parameter grouping, where a group consists of a single parameter.

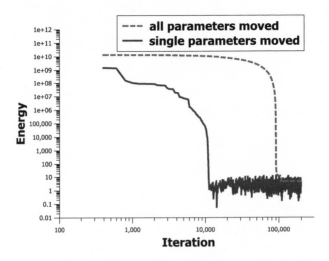

Figure 3—Energy versus iteration number for narrow 6-dimensional Gaussian.

Mixed Δ is used as before, half the time small and half the time $\Delta = 1$. The chain is started out at $\theta = 1$ for all parameters. Without parameter grouping, where all six parameters are simultaneously moved at each iteration, the time to reach equilibration is increased and the acceptance fraction after equilibration is decreased. So, it seems from this example that parameter grouping should always be used; however, there are pitfalls here too, as will be discussed.

These results for the 6-dimensional Gaussian are quite amazing because the region of interest is now a fraction 10^{-36} of the entire volume!

Note that the MCMC algorithm is navigating the chain toward the peak of the steady state distribution using the information in the far tail of the distribution. The reason moving single parameters is more effective than moving all parameters together is that it is easier to find a peak in 1 dimension rather than 6 dimensions.

The distribution is very small in the far tail, and if the distribution tail is truncated to a very small constant value outside of the main support region, which is tiny in these examples, we return to the situation of naïve Monte Carlo. Even worse than a level plane, would be a misleading topology with local maxima.

Multiple candidate algorithm. Table 1 shows the average and standard deviation of θ calculated from the average values of the first and second moments of θ for the narrow 6-dimensional Gaussian.

$$average \equiv Avg(\theta)$$

$$SD = \sqrt{Avg(\theta^2) - \left(Avg(\theta)\right)^2} \quad ,$$

where $Avg(.)$ denotes the average value. The MRT and MC algorithms with different numbers l of candidates per iteration are used, with mixed Δ (15σ and 1) and parameter grouping. The chain is started at $\theta = 1$. The initial 10% of iterations ("burnin" fraction) are ignored in the average value calculation.

Table 1—*Average and standard deviation of a very narrow 6-dimension Gaussian centered at* $\theta = 0.5$ *calculated using different MCMC algorithms. The quantity* l *is the number of candidates considered at each iteration.*

algorithm	iterations	dim	average	SD(10^{-6})
MRT(l=1)	200000	1	0.5	1.07
		2	0.5	1.01
		3	0.5	1
		4	0.5	0.98
		5	0.5	1
		6	0.5	0.99
	100000	1	0.5	1.02
		2	0.5	1
		3	0.5	0.93
		4	0.5	0.96
		5	0.5	1.02
		6	0.5	**12.5**
MRT(l=1000)	200000	1	0.5	0.98
		2	0.5	1.02
		3	0.5	1.02
		4	0.5	0.99
		5	0.5	1.03
		6	0.5	0.98
	100000	1	0.5	0.96
		2	0.5	1
		3	0.5	1
		4	0.5	0.98
		5	0.5	**246.6**
		6	0.5	1.03
MC(l=1000)	2000	1	0.5	1.04
		2	0.5	0.98
		3	0.5	1.1

			4	0.5	1
			5	0.5	1.02
			6	0.5	1.06
		1000	1	0.5	1.04
			2	0.5	1.04
			3	0.5	1.13
			4	0.5	0.87
			5	0.5	1.04
			6	0.5	**23.39**

Table 1 shows that the MRT algorithms reproduce the correct SD with 200 thousand iterations, whether or not multiple candidates are used. The MC-1000 algorithm reproduces the correct SD with 2 thousand iterations. One sees that having multiple candidates is not advantageous with the MRT algorithm, because the same total number of iterations are required. However, for this example using the MC-1000 algorithm with 1000 candidates, the number of chain iterations required are reduced by a factor of 100.

The versions of the MCMC algorithm with multiple candidates are advantageous because they allow parallel processing. The energy calculations for the multiple candidates can be carried out simultaneously in parallel.

Pitfalls associated with random-walk step size and the use of parameter grouping. The simplest rules of thumb gleaned from the above examples would have it that parameter grouping should always be used, and Δ should be started out at 1 and then decreased during an initial "adaptive" phase until a desired acceptance fraction is obtained. To illustrate potential problems with these rules, a multimodal example is used, with

$$\chi^2 = \left(\frac{(\theta - \theta_0 - \Delta_\theta)(\theta - \theta_0 + \Delta_\theta)}{2\Delta_\theta \sigma} \right)^2 , \qquad (1)$$

which has two minima of width σ separated by $2\Delta_\theta$. Figure 4 shows a calculation of the type done in Chapter 9 for an integer-valued chain with $\sigma = 2$, $2\Delta_\theta = 40$, and $l = 10$.

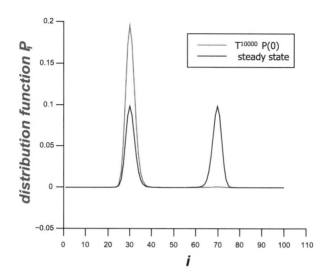

Figure 4—Distribution function after 10000 iterations and the steady state for a bimodal distribution. The time to reach steady state (3τ) is 4×10^6 iterations. The MRT algorithm is used with diagonal band width $l = 10$.

Recall that the diagonal band width l in the integer-valued chain from Chapter 9 corresponds both to 2Δ and to the number of candidates in the continuous parameter case. Figure 4 shows a very long time to reach a steady state, because Δ is not large enough to connect the two modes.

This example shows that the random-walk step size needs to be large enough to connect all modes. Acceptance fraction is not a good guide.

A two-dimensional example relating to parameter grouping has

$$\chi^2 = \left(\frac{(u - \Delta_u)(u + \Delta_u)}{2\Delta_u \sigma}\right)^2 + \left(\frac{v}{\sigma}\right)^2 \quad , \tag{2}$$

where

$$u = \frac{(\theta_1 - \theta_{10}) - (\theta_2 - \theta_{20})}{\sqrt{2}}$$

$$v = \frac{(\theta_1 - \theta_{10}) + (\theta_2 - \theta_{20})}{\sqrt{2}} \quad .$$

This is again a bimodal distribution with two minima of width σ separated by $2\Delta_u$, but rotated around the point $(\theta_{10}, \theta_{20})$, as shown in Fig. 5.

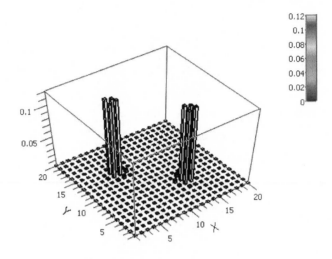

Figure 5—Distribution function for bimodal energy with $\Delta_u = 5$, $\sigma = 0.5$, using an integer-valued simulation of two continuous parameters.

Table 2 shows the time to reach steady state (3τ) for $l = 20$.

Table 2—Time to reach steady state for bimodal energy.

algorithm	grouping?	3τ
MRT	yes	1.5×10^7
MRT	no	212
MC	yes	3.3×10^6
MC	no	2.9

Note the very long times with parameter grouping. For the chain to move from one mode to the other with parameter grouping, where only one parameter is moved at a time, requires an unlikely move to a region of exponentially low probability in one dimension followed by a move to the other mode in the other dimension. If both parameters are moved together such a move can happen with much greater probability.

The continuous version of this calculation in two dimensions is shown in Fig. 6. The two MCMC runs differ in having different starting points (θ's all 0 or 1) and different random number seeds.

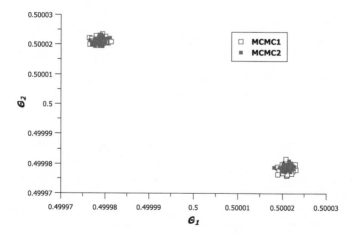

Figure 6—Posterior distribution when Δ is just large enough to connect the two modes. The acceptance fraction is 0.2%.

The example shown in Fig. 6 has the standard deviation of an individual mode $\sigma = 10^{-6}$, the mode separation $2\Delta_\theta = 60 \times 10^{-6}$, and the maximum random step half width $\Delta = 60 \times 10^{-6}$, just large enough to connect the two modes. The MRT algorithm is used, and the acceptance fraction is 0.2 %. For $\Delta = 2\sigma$ (30 times smaller), the acceptance fraction is 46%, but the chain remains stuck at one mode or the other. The same stuckness happens for large Δ when the parameters are moved individually (grouped).

The general rule is that random-walk parameter movement should allow the chain to step across the modal region without having to go through regions of exponentially small probability. Thus it can be dangerous to have Δ too small or parameters in separate groups in some cases.

Table 2 shows that the MC algorithm with $l = 20$ reaches steady state in only 3 iterations and would in this case be very accurate because the random-walk region encompasses almost the entire space.

For this problem with large Δ, the continuous MC algorithm gives satisfactory results in only 100 chain iterations (about the minimum for a reasonable posterior sample) as long as the number of candidates exceeds about 1000.

Discretization. This technique, which will be used later on, changes the problem with continuous variables into a problem involving a single discrete integer-valued variable. This allows parallelization, because the χ^2 (energy) function is first calculated over the entire (discretized) space. The discretized forward model involves functional dependences on various parameters that need to be captured in interpolation tables. This requires a large amount of storage and uses fast interpolation techniques. All of these requirements are well aligned with the future direction of computer technology.

As an example, we use the χ^2 function of Eq. (1), however now mapped into a single discrete index. If the indexing is regular, the energy has two separated modes as shown in Fig. 4. If the indexing is a random permutation, the energy has a random, irregular structure as a function of this index as shown in Fig. 7.

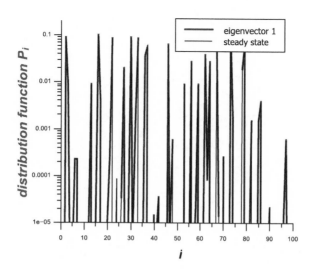

Figure 7—The bimodal energy shown in Fig. 4 mapped into a random one-dimensional index from 1 to 100.

The times to reach steady state for this energy function are shown in Table 3.

Table 3—Times to reach steady state for randomized energy and original bimodal energy.

algorithm	Δ	3τ (rand)	3τ (orig)
MRT	0.05	7×10^7	4×10^6
MRT	0.2	260	7×10^4
MC	0.2	40	6×10^5
MRT	1	27.6	27.6
MC	1	0.085	0.085

The quantity Δ in Table 3 is the half width of the diagonal band of the transition matrix $(l/2)$ divided by the matrix size (100). The MC algorithm with $\Delta = 1$ is generating independent samples from the posterior distribution so the largest eigenvalue is 1 and all the rest are very small. It is interesting that randomization makes no difference for $\Delta = 1$. Note the long times with small Δ, ameliorated to some extent by randomization. With randomization for $\Delta = 0.2$, the MC algorithm is converging about 7 times more rapidly than the MRT, however the steady state (after unscrambling) has some raggedness as shown in Fig. 8.

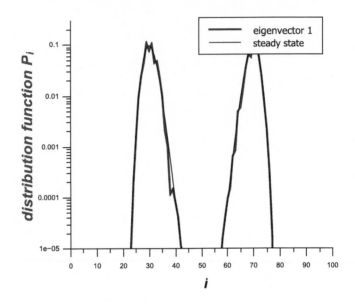

Figure 8—Steady states of the MRT and MC algorithms for randomized energy with $\Delta = 0.2$, after unrandomizing.

Exercises

1. Using a spreadsheet, calculate χ^2 from Eq. (1) on a 50 point mesh.

2. Using a spreadsheet, calculate χ^2 from Eq. (2) on a 20×20 point mesh.

3. Repeat the calculations for Fig. 6 using $\Delta = 2 \times 10^{-6}$ with the Fortran program MCMC provided in the supplementary material for this chapter.

4. To show the correspondence of the MC algorithm to Gibbs sampling, repeat the Fig. 6 calculation using the MC algorithm. Show that the acceptance fraction is 1 when the number of candidates is made large. Also show that the MC algorithm and therefore Gibbs sampling converge extremely slowly when the parameters are moved individually (2 groups of 1 parameter) rather than all together (1 group of 2 parameters).

Chapter 12. Least-Squares Algorithm

We have seen that by suitably defining $\chi(\psi)$ and incorporating the prior through $P(\psi)d\psi = d\theta$, the posterior probability of model parameters given the data can be written as

$$P(\theta \mid Y) \propto e^{-\chi^2(\theta)/2} \quad .$$

The fundamental quantity $\chi^2(\theta)$ is given by

$$\chi^2(\theta) \equiv \sum_{i=1}^{m} \chi_i^2(\theta)$$

in terms of the "scaled residuals" $\chi_i(\theta)$ for each independent measurement. The maximum of the posterior probability occurs at the minimum of the sum of squares given by $\chi^2(\theta)$, and in this chapter a method for finding this minimum is discussed.

Because the methods discussed in this Chapter involve unconstrained model parameters, which we denote by x, and the θ variables are constrained, $0 < \theta < 1$, we use the nonlinear transformations

$$\theta = \left(1 + \exp(-x)\right)^{-1}$$
$$x = -\log(\theta^{-1} - 1) \quad .$$

In this system $-\infty < x < \infty$ and $0 < \theta < 1$ are dimensionless parameters. The physical parameters $X(\theta)$ obtained from the priors have dimensions and constraints.

A case of interest is when the model can be approximated as being linear. In general the residual $\chi_i(x)$ can be represented as a Taylor expansion around some starting point x_0 in parameter space; that is,

$$\chi_i(x) \cong \chi_i(x_0) + \sum_{p=1}^{n} \frac{\partial \chi_i(x_0)}{\partial x_p}(x_p - x_{0,p})$$

$$+ \frac{1}{2}\sum_{p,q=1}^{n} \frac{\partial^2 \chi_i(x_0)}{\partial x_p \partial x_q}(x_p - x_{0,p})(x_q - x_{0,q}) + \dots$$

In a sufficiently small neighborhood of x_0, $\chi(x)$ can be well approximated as linear, depending linearly on parameters x_p for $p = 1,\dots n$, with coefficients given by the Jacobean matrix

$$J_{ip} = \frac{\partial \chi_i}{\partial x_p}(x_0) \quad .$$

Neglecting the second derivatives, which is justified if the problem is not too nonlinear, the fundamental quantity $\chi^2(x)$ is then given by

$$\chi^2(x) = \sum_{i=1}^{m}\left(\chi_i(x_0) + \sum_{p=1}^{n} J_{ip}(x_p - x_{0,p}) \right)^2 \quad .$$

To simplify the notation,

$$x_p - x_{0,p} \rightarrow x_p \quad .$$

The more general form can be restored at the end.

128

Also, to simplify the algebraic manipulations, matrix notation will be used, where, depending on the symbol, quantities may be matrices. For example,

$$A \leftrightarrow \begin{bmatrix} A_{11}, A_{12}, \ldots \\ A_{21}, A_{22}, \ldots \\ \ldots \end{bmatrix} \quad .$$

In terms of matrices,

$$\chi^2(x) = (\chi(0) + Jx)^T (\chi(0) + Jx)$$
$$= \chi^T(0)\chi(0) + x^T J^T \chi(0) + \chi(0)Jx + x^T J^T Jx \qquad (1)$$

where A^T denotes the transpose of matrix A (rows and columns interchanged). One can verify the following needed properties of the transpose,

$$A^T_{i,j} \equiv A_{j,i}$$
$$(AB)^T = B^T A^T \qquad ,$$

and for a square matrix,

$$(A^{-1})^T = (A^T)^{-1} \quad .$$

In Eq. (1), the matrix J is m (the number of data) rows high by n (the number of parameters) columns wide, x is a single column with n rows, and Jx and $\chi(0)$ are single columns with m rows. In matrix multiplication, for example AB, the number of columns of A must be the same as the number of rows of B. The product has the number of rows of A and the number of columns of B.

$$(AB)_{i,j} \equiv \sum_k A_{i,k} B_{k,j} \quad .$$

Our approach to Eq. (1) is to complete the square—a simple but powerful algebraic technique. For example, the quadratic form $ax^2 + bx + c$ can be rewritten as $a(x + b/(2a))^2 + c - b^2/(4a)$, which shows immediately the minimum obtained for $x = -b/(2a)$. Similarly, one can verify that Eq. (1) can be rewritten as

$$\chi^2(x) = (x + (J^T J)^{-1} J^T \chi(0))^T J^T J (x + (J^T J)^{-1} J^T \chi(0))$$
$$+ \chi(0)^T \chi(0) - \chi(0)^T J (J^T J)^{-1} J^T \chi(0)$$

,

or, defining some secondary quantities,

$$\chi^2(x) = (x + H^{-1} g)^T H (x + H^{-1} g) + \chi(0)^T \chi(0) - g^T H^{-1} g$$

, (2)

where the "Hessian" $H \equiv J^T J$ and the "gradient" $g \equiv J^T \chi(0)$. One can verify that the Hessian is symmetric, $H^T = H$. One can also verify that the "gradient" is $1/2$ times the gradient of χ^2 at the point $x = 0$. It is assumed that the $n \times n$ matrix H is nonsingular and can be inverted. It can always be inverted by augmenting the diagonal slightly (even for $m < n$), a technique that will be discussed subsequently. When $m = n$, if the $m \times m$ matrix J itself is invertible,

$$H^{-1} g = J^{-1} (J^T)^{-1} J^T \chi(0) = J^{-1} \chi(0) \quad .$$

To further elucidate Eq. (2), we introduce a change of the "coordinate system," with old coordinates related to new coordinates by the $m \times m$ linear transformation matrix L,

$$\chi = L \tilde{\chi} \quad ,$$

(3)

130

The transformation matrix L in Eq. (3) is constructed as

$$L = [J \mid O] \quad ,$$

that is, L is the juxtaposition of the m row by n column matrix J with a new orthonormal matrix O consisting of m rows and $m-n$ columns, together making up an $m \times m$ square matrix. The orthonormal matrix O has columns that are orthogonal to every other column of L, those of J as well as those of O itself.

One can verify that

$$L^{-1} = \begin{bmatrix} H^{-1}J^{T} \\ O^{T} \end{bmatrix} \tag{4}$$

satisfies $L^{-1}L = I$. Because an invertible matrix has a unique inverse, it is also a right-side inverse as well as a left-side inverse.

In terms of the new coordinates,

$$g = J^{T}L\tilde{\chi}(0) = [H, 0]\tilde{\chi}(0) \quad ,$$

so that the n components of $H^{-1}g$ are $\tilde{\chi}_{p}(0)$, $p = 1,...n$. In fact, one sees that

$$\chi^{2}(x) = \sum_{p,q=1}^{n} H_{p,q}(x_{p} + \tilde{\chi}_{p}(0))(x_{q} + \tilde{\chi}_{p}(0)) + \sum_{j=n+1}^{m}(\tilde{\chi}_{j}(0))^{2} \quad . \tag{5}$$

Equation (5) is the desired form. It shows that the $\chi^{2}(x)$ attains a minimum for the parameter values

$$x_p = -\tilde{\chi}_p(0) = -(L^{-1}\chi(0))_p = -(H^{-1}J^T\chi(0))_p = -(H^{-1}g)_p$$

and it shows the minimum so obtained, which can also be written as

$$\chi^2_{min} = \chi^T(0)(I-P)\chi(0) \quad ,$$

in terms of the projection operator $P \equiv JH^{-1}J^T$ (see Exercise 1).

If $n \geq m$, the minimum $\chi^2(x)$ can be reduced to zero.

Two-parameter example. In this example there are 3 data points with scaled residuals $\chi_1(0)$, $\chi_2(0)$, $\chi_3(0)$ and 2 parameters x_1 and x_2.

$$\chi(x) = \begin{bmatrix} \chi_1(x) \\ \chi_2(x) \\ \chi_2(x) \end{bmatrix} = \begin{bmatrix} \chi_1(0)-x_1 \\ \chi_2(0)-x_1-2x_2 \\ \chi_3(0)-x_1+x_2 \end{bmatrix} \quad .$$

Therefore

$$J = \begin{bmatrix} -1 & 0 \\ -1 & -2 \\ -1 & 1 \end{bmatrix}$$

and

$$H = J^TJ = \begin{bmatrix} -1 & -1 & -1 \\ 0 & -2 & 1 \end{bmatrix}\begin{bmatrix} -1 & 0 \\ -1 & -2 \\ -1 & 1 \end{bmatrix} = \begin{bmatrix} 3 & 1 \\ 1 & 5 \end{bmatrix}$$

with inverse

$$H^{-1} = \frac{1}{14}\begin{bmatrix} 5 & -1 \\ -1 & 3 \end{bmatrix} .$$

The transformation matrix L is given by

$$L = [J \,|\, O] = \begin{bmatrix} -1 & 0 & O_1 \\ -1 & -2 & O_2 \\ -1 & 1 & O_3 \end{bmatrix}$$

with the orthonormal vector O_1, O_2, and O_3 determined from the equations

$$-O_1 - O_2 - O_3 = 0$$
$$-2O_2 + O_3 = 0$$
$$O_1^2 + O_2^2 + O_2^2 = 1$$

having the solution

$$O = \frac{1}{\sqrt{14}}\begin{bmatrix} -3 \\ 1 \\ 2 \end{bmatrix} .$$

The transformed residuals are given by

$$\tilde{\chi} = L^{-1}\chi = \begin{bmatrix} H^{-1}J^T \\ O^T \end{bmatrix}\chi$$

and in terms of transformed residuals

$$\chi(x)^2 = \frac{1}{14}\left(3(x_1 + \tilde{\chi}_1(0))^2 + 2(x_1 + \tilde{\chi}_1(0))(x_2 + \tilde{\chi}_2(0)) + 5(x_2 + \tilde{\chi}_2(0))^2\right) + (\tilde{\chi}_3(0))^2 .$$

Because

$$H^{-1}J^T = \frac{1}{14}\begin{bmatrix} 5 & -1 \\ -1 & 3 \end{bmatrix}\begin{bmatrix} -1 & -1 & -1 \\ 0 & -2 & 1 \end{bmatrix} = \frac{1}{14}\begin{bmatrix} -5 & -3 & -6 \\ 1 & -5 & 4 \end{bmatrix} ,$$

the transformed scaled data are given by

$$\tilde{\chi}_1(0) = \frac{1}{14}\left(-5\chi_1(0) - 3\chi_2(0) - 6\chi_3(0)\right)$$

$$\tilde{\chi}_2(0) = \frac{1}{14}\left(\chi_1(0) - 5\chi_2(0) + 4\chi_3(0)\right)$$

$$\tilde{\chi}_3(0) = \frac{1}{\sqrt{14}}\left(-3\chi_1(0) + \chi_2(0) + 2\chi_3(0)\right)$$

One-parameter example. In this example there are 3 data points and 1 parameter x.

$$\chi^2(x) = (Y_1 - xf_1)^2 + (Y_2 - xf_2)^2 + (Y_3 - xf_3)^2$$

One can verify that

$$J = \begin{bmatrix} -f_1 \\ -f_2 \\ -f_3 \end{bmatrix}$$

$$H = J^T J = f_1^2 + f_2^2 + f_3^2 \equiv f^2$$

$$H^{-1} = \frac{1}{f^2}$$

The first column of L is

$$L = \begin{bmatrix} -f_1, \ldots \\ -f_2, \ldots \\ -f_3, \ldots \end{bmatrix}$$

The orthogonal matrix O forming the next 2 columns of L can be constructed in an infinite number of ways. One solution is to choose by inspection a first column with only 2 nonzero elements orthogonal to the first column of L and then solve 2 linear equations for the elements of the 3rd column expressing orthogonality to the 1st and 2nd columns. The final step is to normalize the columns. The result of this is

$$
O = \begin{bmatrix} \dfrac{f_2}{d}, & \dfrac{f_1 f_3}{fd} \\[2ex] -\dfrac{f_1}{d}, & \dfrac{f_2 f_3}{fd} \\[2ex] 0, & -\dfrac{f_1^2 + f_2^2}{fd} \end{bmatrix} \quad,
$$

where

$$
d = \sqrt{f_1^2 + f_2^2}
$$

and $f = \sqrt{f_1^2 + f_2^2 + f_3^2}$ as defined above.

The final expression is then

$$
\chi^2(x) = f^2(\tilde{Y}_1 - x)^2 + (\tilde{Y}_2)^2 + (\tilde{Y}_3)^2 \quad,
$$

with

$$\widetilde{Y}_1 = \frac{Y_1 f_1 + Y_2 f_2 + Y_3 f_3}{f^2}$$

$$\widetilde{Y}_2 = \frac{Y_1 f_2 - Y_2 f_1}{d}$$

$$\widetilde{Y}_3 = \frac{Y_1 f_1 f_3 + Y_2 f_2 f_3 - Y_3 d^2}{df}$$

As a special case of the forgoing example, let $f_1 = f_2 = f_3 = 1$. The fit parameter at the minimum $x = \widetilde{Y}_1$ is then the mean of the 3 data points. If the data come from uncorrelated Gaussian distributions (with the same mean and standard deviation 1), the average value of χ^2_{min} is 2, 1 being contributed by each of the $m - n = 3 - 1 = 2$ degrees of freedom,

$$\left\langle (\widetilde{Y}_2)^2 \right\rangle = \frac{\left\langle (Y_1)^2 \right\rangle f_1^2 + \left\langle (Y_2)^2 \right\rangle f_2^2}{d^2} = 1$$

$$\left\langle (\widetilde{Y}_3)^2 \right\rangle = \frac{\left\langle (Y_1)^2 \right\rangle f_1^2 f_3^2 + \left\langle (Y_2)^2 \right\rangle f_2^2 f_3^2 + \left\langle (Y_3)^2 \right\rangle d^4}{d^2 f^2} = 1$$

Denominator in formula for variance. We are now in a position to understand $n - 1$ rather than n in the denominator of Eq. (1) of Chapter 1. Consider a set of independent random variables Y_i, each with the same variance and mean. We linearly transform the sum of squares by using the new variables found above,

$$\widetilde{Y}_1 = \left\langle Y \right\rangle_m \equiv \frac{1}{m} \sum_{i=1}^{m} Y_i$$

and $\widetilde{Y}_2 = (Y_1 - Y_2)/\sqrt{2}$, $\widetilde{Y}_3 = (Y_1 + Y_2 - 2Y_3)/\sqrt{6}$, ... as above. Therefore,

$$\sum_{i=1}^{m} (Y_i - x)^2 = m(\tilde{Y}_1 - x)^2 + \sum_{i \geq 2}^{m} (\tilde{Y}_i)^2 \quad .$$ (6)

For the new variables except the first, $\left\langle (\tilde{Y}_i)^2 \right\rangle = Var(Y)$. Letting $x = \tilde{Y}_1 = \left\langle Y \right\rangle_m$ and taking the average of Eq. (6), we arrive at the formula

$$\left\langle \sum_{i=1}^{m} (Y_i - \langle Y \rangle_m)^2 \right\rangle = (m-1)Var(Y) \quad .$$

Nonlinear least squares. The formalism can be extended to a general nonlinear minimization scheme. For a linear problem, the minimum is immediately calculated as a step

$$s = -H^{-1}g$$

from the current point in parameter space.

For a nonlinear problem, this step (Newton's step) may be too ambitious and may not decrease χ^2. Because the gradient of $f \equiv \chi^2$ near the origin is $2g$, a step of the form

$$s = -\frac{1}{\mu} g$$

with μ sufficiently large will be a step down the gradient (steepest descent) and will always decrease f. The disadvantage of this conservative step (Gaussian step) is that it may take a very large number of steps to reach the minimum, while the Newton step goes immediately to the minimum for a linear problem and gives quadratic convergence near the minimum (the error eventually decreasing like the square of some factor at each step). The Levenberg-Marquardt [1,2] compromise solution is a step of the form

$$s = -(H + \mu I)^{-1} g \qquad (7)$$

that becomes the Newton step for small μ and is a Gaussian step for large μ. This also takes care of possible singularity of H, because the augmented matrix $H + \mu I$ will always be invertible for sufficiently large μ (the example of Exercise 2 may be helpful).

To understand the Levenberg-Marquardt method, note that Eq. (7), for some $\mu > 0$ is the solution of the constrained minimization of the quadratic form given by Eq. (2) subject to the constraint on step size (see Exercise 3)

$$\sqrt{s^T s} \leq \delta \quad . \qquad (8)$$

Therefore, the solution of Eq. (7) represents the minimum of Eq. (2) within a "trust radius" δ given by Eq. (8).

Nonlinear minimization algorithms of this type use various schemes for choosing the value of μ; for example, the general methods discussed by Dennis and Schnabel and the particular scheme employed by Klare in his Fortran program GN (Gauss Newton least-squares minimization), which works well and will be briefly described here. The Fortran for this algorithm is included in the supplementary material for this chapter.

Let $\Phi(\mu)$ be the magnitude of the step from Eq. (8) minus δ. In order to find the reduced step we need to solve the equation

$$\Phi(\mu) \equiv \sqrt{s^T s} - \delta = 0 \quad , \qquad (9)$$

for μ and then compute the step from Eq. (7). Given δ, Eq. (9) is solved iteratively for μ (see Exercises 4–8).

The GN algorithm by Klare [4] proceeds as follows:

First, try a full Newton step, with μ just the minimum required to prevent singularity. If the problem is linear, the minimum will have been obtained in this one step.

If the function increases, try a smaller step, the "Cauchy step" (see Exercise 9), but still in the Newton direction. The Cauchy step provides a step length, without requiring matrix inversion, in the fallback mode where the Newton step does not decrease f.

After that, one has available an initial f_0, a final f if the step were taken, and the derivatives at the initial point. The trust radius is then revised using a quadratic fit based on the initial and final values of f and the initial derivative (see Exercise 10). The revised trust radius is used to determine a new value of μ and a new step from Eq. (7).

For example, consider a test problem proposed by Rosenbrock[4], which has $m = n = 2$. The residuals are defined as

$$\chi(x) = \begin{pmatrix} 10(x_2 - x_1^2) \\ 1 - x_1 \end{pmatrix} . \tag{10}$$

Figure 1 shows the steps to the minimum taken by GN for two different starting points.

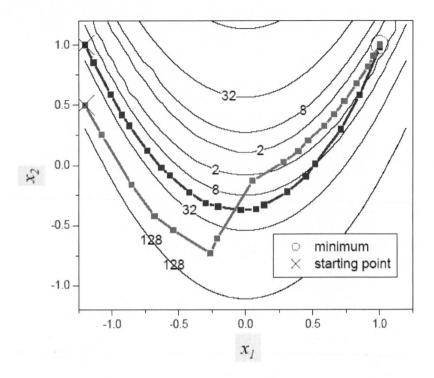

Figure 1—Steps taken to reach the minimum superimposed on a contour plot of the function $f(x) = \chi(x)^T \chi(x) \equiv \chi^2(x)$ for the Rosenbrock test problem. The number of function evaluations required was 39 or 36 depending on the starting point.

It is interesting to note in Fig. 1 that the Newton step is almost at right angles to the steepest descent direction, and stays high until reaching the region of quadratic convergence at the end.

This method of nonlinear least-squares minimization (maximum likelihood) usually is satisfactory and requires many times less forward calculations than the MCMC techniques discussed in this book. It is an example of the "old" way of approaching the problem of interpretation of data. Given a dataset and a starting point in parameter space, by minimizing $\chi^2(x)$ one obtains a single "best fit" value of the model parameters.

If the number of parameters is large, it is necessary to group them (as in MCMC) and iteratively minimize, moving one group at a time.

Parameter uncertainty. For a linear problem parameter uncertainty is estimated by generating alternate datasets (alternate realizations of the data) by adding uncorrelated Gaussian random numbers with mean 0 and standard deviation 1 to the residuals. One then repeats the least-squares minimization for the alternate datasets to find alternate values of the parameters. For linear problems a linearized analysis can be used where the variations in the parameters are obtained just by multiplying the data variations by the "error" matrix

$$e_{p,i} = \frac{\partial x_p}{\partial \chi_i} = (H^{-1}J^T)_{p,i} \quad .$$

Assuming uncorrelated data with average variations given by

$$\langle (\Delta \chi_i)(\Delta \chi_j) \rangle = \delta_{i,j} \quad ,$$

where $\Delta \chi_i$ denotes the variation of data point i, and the Kronecker delta $\delta_{i,j}$ is 1 if the two indices match, otherwise zero, the covariance matrix is given by the inverse of the Hessian,

$$\langle (\Delta x_p)(\Delta x_q) \rangle = \left(ee^T \right)_{p,q} = (H^{-1}J^T J (H^{-1})^T)_{p,q} = (H^{-1})_{p,q} \quad .$$

Figure 2 shows these parameter uncertainty estimates for the Rosenbrock test problem versus the correct uncertainties obtained by generating alternate possible interpretations of the actual data from the posterior distribution $\exp(-\chi^2(x)/2)$. It is interesting that even in this nonlinear case, generating alternate realizations of the data and repeating the fit seems to give the correct answer. For a nonlinear problem the linearized uncertainty analysis often does not agree with the nonlinear analysis, although it is exact for a linear problem. It becomes exact also for a nonlinear problem in the limit of very small uncertainties.

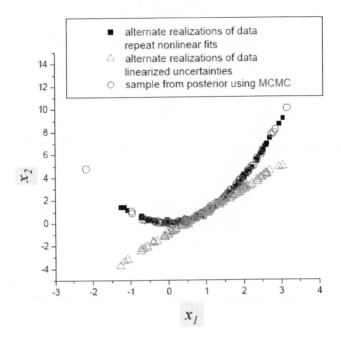

Figure 2—Parameter uncertainty for the Rosenblock test problem by generating 100 alternate realizations of the data by adding uncorrelated Gaussian random numbers with standard deviation 1 to the residuals. The result is in good agreement with a sample from the posterior $\exp(-\chi^2(x)/2)$ *obtained using MCMC.*

For this problem the parameter uncertainties are almost one-dimensional because of the narrow, flat-bottomed-valley character of the $\chi^2(x)$ function.

Generating alternate realizations of the data and repeating the nonlinear minimization of $\chi^2(x)$ is not in general the same as generating from the posterior $\exp(-\chi^2(x)/2)$. A simple counterexample is given in Exercise 11.

When the computation time is not an issue, as more and more is the case, the MCMC approach is preferred over minimization of $\chi^2(x)$ because of its conceptual simplicity and correctness. Nonlinear least-squares minimization still has a place in relatively quickly moving the parameters into the region where the posterior probability is large.

Exercises

1. Show that the matrix $P \equiv JH^{-1}J^T$ satisfies $P^2 = P$ and $PJ = J$. Thus P is a projection operator that acts as an identity operator in an n-dimensional subspace of the entire m-dimensional space spanned by the columns of the matrix J. Similarly, the operator $I - P$, where I is the $m \times m$ identity matrix (all 1's on the diagonal), is a projection operator on the $(m - n)$-dimensional subspace that is the null subspace of P.

2. Consider the case where there are two parameters and one residual r. Write the equations for f, J, H, and g. Write out the equation for step s from $(H + \mu I)s = -g$ and solve for s. Substitute for g to

obtain $s = -\dfrac{\left(\begin{array}{c} \dfrac{\partial r}{\partial x_1} \\ \dfrac{\partial r}{\partial x_2} \end{array} \right)}{\left(\left(\dfrac{\partial r}{\partial x_1} \right)^2 + \left(\dfrac{\partial r}{\partial x_2} \right)^2 \right) + \mu} r$. Thus there is a well-

defined step directed down the gradient in the limit $\mu \to 0$, even though the matrix H is singular.

3. Show that the minimization of Eq. (1) is equivalent to the minimization problem: $2 \sum_p g_p s_p + \sum_{p,q} H_{p,q} s_p s_q = \min$, where s is the step, H is the Hessian, and g is the gradient. Now include the trust region constraint $\sum_p (s_p)^2 = \delta^2$, and using a Lagrange multiplier show that the constrained minimum satisfies the equation $s(\mu) = -(H + \mu I)^{-1} g$ for some $\mu > 0$.

4. Motivated by the one-dimensional case, Dennis and Schnabel suggest a local model for $\Phi(\mu)$,

$$\Phi(\mu) \cong \frac{\alpha}{\beta + \mu} - \delta \quad,$$

for some scalar parameters α and β. Assuming a current value μ_c of μ, find the current values of α_c and β_c using $\Phi(\mu_c)$ and $\Phi'(\mu_c)$.

5. Show that the iteration of μ based on the local model is

$$\mu_+ = \frac{\alpha_c}{\delta_c} - \beta_c \quad.$$

6. Using the result of the previous exercise, show that

$$\mu_+ = \mu_c - \frac{\Phi(\mu_c) + \delta_c}{\delta_c} \frac{\Phi(\mu_c)}{\Phi'(\mu_c)} \quad.$$

This equation is used to find an iterative solution of $\Phi(\mu) = 0$. What is the iteration using Newton's method?

7. Derive the derivative with respect to μ of the column matrix $s(\mu)$ defined by $s(\mu) = (H + \mu I)^{-1} g$. Hint: Take the derivative of the equation $(H + \mu I) s(\mu) = g$ with respect to μ and then solve for $s'(\mu)$.

8. With $s(\mu)$ as in the previous exercise and
$\Phi(\mu) = \sqrt{s(\mu)^T s(\mu)} - \delta$, derive the

expression: $\Phi'(\mu) = -\dfrac{s(\mu)^T (H + \mu I)^{-1} s(\mu)}{\sqrt{s(\mu)^T s(\mu)}} \quad.$

9. The Cauchy step is the step in the direction g that minimizes the quadratic given by Eq. (2). Show that the magnitude of this step is given by

$$s_C = \frac{(g^T g)^{3/2}}{g^T H g} \quad.$$

10. Consider movement in the direction of the last step s, $x = \lambda s / \| s \|$ and the value of $f(\lambda)$ as a function of λ. Show that the minimum occurs at

$$\lambda_* = \frac{1}{2} \frac{\| s \|}{1-r} \quad,$$

with

$$r = \frac{f_0 - f}{-2g^T s} \quad.$$

The quantity r is bounded above by 1 (when the curvature is negligible so that $f \cong f_0 + 2g^T s$) and unbounded below (when $f > f_0$, $g^T s$ is negative). With the revised trust radius given by λ_*, which can be greater than the last step if $r > 1/2$, a new value of μ and a new step are calculated.

11. Let $\chi(\psi) = (\psi - \psi_0)/\sigma$, with $\psi(\theta) = \theta^{1/p}$. Compare generating x from $\exp(-\chi^2(x)/2)$ with generating alternate values of data ψ_0 and then finding the minimum-χ^2 values of x.

References

1. Levenberg, K. "A Method for the Solution of Certain Problems in Least Squares." *Quart. Appl. Math.* 2:164–168 (1944).
2. Marquardt, D. "An Algorithm for Least-Squares Estimation of Nonlinear Parameters." *SIAM J. Appl. Math.* 11:431-441 (1963).
3. Dennis, Jr., J. E., and Robert B. Schnabel. *Numerical Methods for Unconstrained Optimization and Nonlinear Equations.* Classics in Applied Mathematics 16, Society for Industrial and Applied Mathematics (1983).
4. Klare, Kenneth, and Guthrie Miller. "GN—A Simple and Effective Nonlinear Least-Squares Algorithm for the Open Source Literature." www.netlib.org/misc/gn/ (2013).

Chapter 13. Self-Consistency of Model and Data

The important question addressed in this chapter is "How can one know that the modeling is consistent with the data and no further effort need be expended to improve the model or recheck the data?" A rather obvious semi quantitative rule will be first stated and later analyzed in some simple cases. The rule, in terms of the posterior average value of $\chi^2(\theta)$, is the following:

$$\frac{\left\langle \chi^2(\theta) \right\rangle}{m} \cong 1 \quad , \tag{1}$$

where m is the number of data points. That is, the posterior average of $\chi^2(\theta)$ should be approximately 1 per data point.

As discussed in Chapter 7, the fundamental quantity is the likelihood function for a single measurement, which is defined as the probability of the measurement result given the model parameters, considered as a function of the model parameters and normalized to give 1 at the maximum likelihood. For m independent measurements, the combined likelihood $L(\theta)$ is the product of the likelihoods for the individual measurements. The quantity $\chi^2(\theta)$ is then defined by the relationship

$$L(\theta) = \exp(-\chi^2(\theta)/2) \quad , \tag{2}$$

which is the same as saying that $\chi^2(\theta)$ is -2 times the log of the likelihood function. The quantity $\chi^2(\theta)$ is given by the sum over all data points of the squares of the scaled residuals, and in cases where the data are approximately normally distributed (for example counting measurements with counts not too small), the scaled residual is the data minus fit divided by the standard deviation of the data, where the fit is the forward calculation of the true value of the measurement from the model parameters. If the model gives a good representation of the data, the scaled residuals should, on average, be 1, as if the data were generated using random numbers starting with the model calculation. Many large residuals indicate a problem where the model is not consistent with the data, as if the data were generated from a different model.

In linear least-squares fitting, for the best fit one expects $\chi^2(\theta)$ to equal about 1 *per degree of freedom*, rather than per data point, where the degrees of freedom are the number of data points minus the number of fit parameters. In fact, in fitting a linear model with the number of parameters equal to or greater than the number of data points, $\chi^2(\theta)$ normally can be reduced to zero. In probabilistic data modeling we are concerned with the posterior average value of $\chi^2(\theta)$ rather than its minimum value.

Analysis of the rule for a linear model. Let us consider this rule in the special case of a linear model. Start with the case where the number of model parameters n is less than or equal to the number of data points m. From Eq. (5) of Chapter 12, assuming constraints do not come into play and therefore replacing the unconstrained x variables of Chapter 12 with the θ variables used in MCMC,

$$\chi^2(\theta) = \sum_{p,q=1}^{n} H_{p,q}(\theta_p - \tilde{\chi}_p(0))(\theta_q - \tilde{\chi}_q(0)) + \sum_{j=n+1}^{m} (\tilde{\chi}_j(0))^2 \quad ,$$

where $H = J^T J$ is symmetric positive definite, and $\tilde{\chi}_j(0)$ represents the scaled residual for $\theta = 0$ in a transformed coordinate system. On average for normally distributed data, each term in the right-hand summation gives 1 for a total of $m - n$. For our simple case of normal data and a uniform prior,

$$\chi_j(\theta) \leftrightarrow \frac{Y_j - f_j(\theta)}{\sigma_j} \quad ,$$

where Y_j, $f_j(\theta)$, and σ_j are data value, fit value, and standard deviation for the j th data point. The residual is the difference between data and fit scaled by the uncertainty of the data, and $\tilde{\chi}_j(0)$ represents the scaled residual for $\theta = 0$ in the transformed coordinate system.

The first summation term, which is a quadratic form involving H, can be transformed to a sum of squares by means of another linear transformation, giving

$$\chi^2(\theta) = \sum_{p=1}^{n} \left(\hat{\theta}_p - \hat{\chi}_p \right)^2 + \sum_{j=n+1}^{m} (\tilde{\chi}_j(0))^2 .$$

Thus we have

$$\frac{\int \exp(-\chi^2/2)\chi^2 d\theta_1 d\theta_2 \ldots d\theta_n}{\int \exp(-\chi^2/2) d\theta_1 d\theta_2 \ldots d\theta_n} = n + \sum_{j=n+1}^{m} (\tilde{\chi}_j(0))^2 \cong m , \qquad (3)$$

obtained by changing variables $\theta_p \to \hat{\theta}_p$ in the n-dimensional integral. The integral of each term $(\hat{\theta}_p - \hat{\chi}_p)^2$ in the sum over p gives 1.

If the number of parameters exceeds the number of data, by a coordinate transformation, $n - m$ parameters can be found that have no effect on $\chi^2(\theta)$. Integration over these parameters has no effect. Integration over the remaining m parameters gives Eq. (1) exactly.

Influence of the prior. To understand the influence of the prior, consider a simple example with 1 data point and 1 parameter, assuming that

$$\chi(\theta) = \frac{Y - \theta}{\sigma} .$$

The prior is assumed to be normal, centered on parameter value θ_0 and having standard deviation σ_0. The integration in Eq. (3) then becomes

$$\left\langle \chi^2(\theta) \right\rangle =$$

$$\frac{\int d\theta \left(\frac{Y-\theta}{\sigma} \right)^2 \exp\left(-\frac{1}{2}\left(\frac{Y-\theta}{\sigma} \right)^2 \right) \exp\left(-\frac{1}{2}\left(\frac{\theta-\theta_0}{\sigma_0} \right)^2 \right)}{\int d\theta \exp\left(-\frac{1}{2}\left(\frac{Y-\theta}{\sigma} \right)^2 \right) \exp\left(-\frac{1}{2}\left(\frac{\theta-\theta_0}{\sigma_0} \right)^2 \right)} \quad . \qquad (4)$$

In Eq. (4), the product of the two Gaussians is again a Gaussian with standard deviation and mean given by

$$\frac{1}{\overline{\sigma}^2} = \frac{1}{\sigma^2} + \frac{1}{\sigma_0^2} \quad .$$

$$\overline{\theta} = \frac{\dfrac{Y}{\sigma^2} + \dfrac{\theta_0}{\sigma_0^2}}{\dfrac{1}{\sigma^2} + \dfrac{1}{\sigma_0^2}}$$

Furthermore,

$$(Y-\theta)^2 = (\theta-\overline{\theta})^2 - 2(\theta-\overline{\theta})(Y-\overline{\theta}) + (Y-\overline{\theta})^2 \quad .$$

If the integration variable is changed to $\theta-\overline{\theta}$, the linear term vanishes, and we obtain

$$\left\langle \chi^2(\theta) \right\rangle = \frac{\overline{\sigma}^2 + (Y-\overline{\theta})^2}{\sigma^2} = \left(\frac{\overline{\sigma}}{\sigma} \right)^2 \left(1 + \left(\frac{\overline{\sigma}}{\sigma_0} \right)^2 \left(\frac{Y-\theta_0}{\sigma_0} \right)^2 \right) \quad .$$

One can think of the dimensionless parameter σ_0 / σ as a measure of the strength of the prior. If the prior is weak, meaning that $\sigma_0 / \sigma \gg 1$, then $\bar{\sigma} \cong \sigma$ and $\langle \chi^2(\theta) \rangle \cong 1$. This is the situation with lots of data, where the data dominate the prior. If the prior is strong, meaning that $\sigma_0 / \sigma \ll 1$, then $\bar{\sigma} \cong \sigma_0$, and

$$\langle \chi^2(\theta) \rangle \cong \left(\frac{Y - \theta_0}{\sigma} \right)^2 \ .$$

This relationship can be understood directly from Eq. (4), because the posterior distribution in this case is concentrated at θ_0 so that $Y - \theta \cong Y - \theta_0$. In this case the prior can substantially either decrease or increase $\chi^2(\theta)$.

Four iterative steps. In practice the uncertainties are often not known very well, and because a two-fold increase of the uncertainties causes a four-fold decrease in $\chi^2(\theta)$, adjustments of the uncertainties has a pronounced effect on $\chi^2(\theta)$. Therefore, Eq. (1) can be regarded as a rule of thumb. The modeling process consists in iterating the four steps:

1) Hypothesize a model structure and calculate $\langle \chi^2(\theta) \rangle / m$. If approximately 1, quit.
2) Provide an independent check of the data and possibly eliminate spurious data points.
3) Consider the effect of the prior in possibly reducing or increasing $\langle \chi^2(\theta) \rangle / m$.
4) Possibly adjust the data uncertainty parameters.

At the end of this process, Eq. (1) should hold at least approximately. Note that if the data uncertainties are increased, this directly affects the parameter uncertainties calculated in the end. And if $\langle \chi^2(\theta) \rangle$ per data point is significantly larger than 1, the parameter uncertainties are in question. The true parameter uncertainties might be larger, as would result from increased data uncertainties to achieve $\langle \chi^2(\theta) \rangle = 1$ per data point.

Therefore iterating these four steps is essential to provide closure on the entire modeling process, even if the data uncertainties are unknown and must be adjusted to satisfy Eq. (1).

Chapter 14. Hypothesis Testing

As a starting point, consider the white-black atom example of Chapter 8 once again. However, now we assume two hypotheses: H_0 where there are no black atoms, only white atoms, and H_1 where there are white and black atoms in the ratio p to q (prior probabilities of white and black). Let us enumerate the possibilities and their probabilities. These are shown in Table 1.

Table 1—Probabilities of the various possibilities for two hypotheses.

hypothesis	atom color	meas0	meas1
0 (only white)	white	$p_0(1-\alpha)$	$p_0\alpha$
1 (white and black)	white	$p_1 p(1-\alpha)$	$p_1 p\alpha$
1 (white and black)	black	$p_1 q\beta$	$p_1 q(1-\beta)$

In Table 1 α and β are parameters of the likelihood function as discussed in Chapter 8—the false positive and false negative rates, and p_0 and p_1 are the prior probabilities of hypotheses 0 and 1. We see immediately that if 1 is measured (nominally indicating a black atom), the probability of hypothesis 0 is given by

$$P(H_0 \mid meas1) = \frac{p_0\alpha}{p_0\alpha + p_1\left(p\alpha + q(1-\beta)\right)} . \tag{1}$$

In this equation α and $p\alpha + q(1-\beta)$ are the sums of the likelihood function when 1 is measured over the respective priors for the two hypotheses. Let us say that the numbers are as before in Chapter 8, $p = 0.999$, $q = 0.001$, $\alpha = \beta = 0.05$, and p_0 and p_1 are taken to be equal. We find the posterior probability of hypothesis 0 is 50%, meaning that hypothesis 0 and hypothesis 1 are equally likely. This is counterintuitive, because meas1 nominally means that the atom color is black. This correct interpretation is a result of the smallness of the prior probability of black atoms, q. If $p = q$, the posterior probability of hypothesis 0 would be a small number, approximately α, the false positive rate.

Hypotheses testing is a comparison of alternate modeling parameterizations in contrast to being a model itself. For example, if we had defined Hypothesis 0 as white only and Hypothesis 1 as black only, hypothesis testing would not be different from the basic white/black model defined in Chapter 8 ($q = 1$, and p_0 and p_1 are equivalent to p and q of Chapter 8, Table 1).

On the other hand, hypothesis is just another forward-model parameter like any other, so, given a prior probability for hypotheses, the rules of conditional probability yield a posterior probability in the usual way. In this book hypothesis testing is used only with equal assumed prior probabilities. Posterior probabilities of hypotheses, given the data, are calculated, and because they are based on new measurements, these posterior probabilities are more reliable than prior probabilities. In the limit of strong data, the new data overwhelm the prior, and the "best" hypothesis is the one with the largest posterior probability.

Evidence supporting competing models. In general in data modeling there are many competing models that one might use. If one were to do separate analyses with several different models, the probability of a particular hypothesis would be proportional to

$$P(H_h \mid data) \propto \int d\xi_h P(\xi_h) L(\xi_h; H_h) \quad , \tag{2}$$

where ξ_h are the model parameters for hypothesis h, $P(\xi_h)$ is the prior probability distribution, and $L(\xi_h; H_h)$ is the likelihood function for hypothesis h. In words, the posterior probability of a particular hypothesis is proportional to the integral, involving all the model parameters for the hypothesis, of the likelihood function times the prior. For a given hypothesis, this integral is the denominator in Bayes theorem, which is not required in an ordinary Markov Chain Monte Carlo (MCMC) analysis. The value of this integral is called the "Evidence" for the hypothesis. The proportionality in Eq. (2) is normalized over the entire "universe of possibilities" for hypothesis h, needing numerical values of the evidence for each hypothesis, which can be calculated one-by-one.

With a dataset that can be split (more than one measurement), we could use as a prior the posterior obtained by conditioning on the first half of the dataset. Then the prior average becomes the posterior average, given the first half of the data, of the combined likelihood function for the second half of the dataset, which, as a posterior average, can be straightforwardly evaluated using Markov Chain Monte Carlo (MCMC).

For a large dataset, half of the dataset is still a large dataset and is equivalent in this sense to the dataset itself. So, using the split dataset argument, the posterior rather than prior average of the likelihood function would seem to give the relative posterior probability of an assumed hypothesis.

When the priors for each parameter in Eq. (2) are independent (the combined prior is the product of individual parameter priors), and the individual priors are incorporated into θ variables as discussed in Chapter 8, the problem of evaluating the evidence for a particular hypothesis is equivalent to the problem of evaluating a multidimensional integral of the form

$$E \equiv \int d\xi\, P(\xi) L(\xi) = \int_0^1 d\theta\, L(\theta) \quad . \tag{3}$$

All dimensions of θ have domain 0 to 1.

The evidence integral depends on the extent of the assumed prior relative to the support region of the likelihood function and can be seen to give a relative measure of how well the prior matches the data. The example considered in Exercise 1 is modeling data that are a function of time with either a constant (1 parameter) or a constant plus linear slope (2 parameters). Introducing the slope parameter changes the 1-parameter evidence integral by multiplication by a factor

$$\frac{E(II)}{E(I)} = \frac{\sqrt{2\pi}\sigma_b\sqrt{1-\gamma^2}}{\Delta_b} \ .$$

The change factor is the ratio of the posterior uncertainty of the slope (σ_b) to the full extent of the uniform prior on the slope (Δ_b), which is a small number for a broad prior, modified by parameter correlation (γ).

If both 1- and 2-parameter models produce good values of χ^2, the simpler, 1-parameter model has the largest evidence integral by an amount that depends on the extent of the prior on the slope.

The evidence integral and parameter uncertainty. The evidence integral would seem closely related to parameter uncertainty. We can write

$$E = L_{max}V$$

where L_{max} is the maximum of the likelihood function and V is a measure of the volume of the support region of the likelihood function compared the entire volume of the prior. A measure of single-parameter uncertainty in terms of θ variables might seem to be $V^{1/n}$.

Because of parameter correlation, which can decrease the volume of the support region of the likelihood function, instead of $V^{1/n}$ a better and easier-to-calculate measure of parameter uncertainty to be used in model selection is

$$U = \left(\prod_{i=1}^{n}\sigma_i\right)^{1/n} , \tag{4}$$

where σ_i is the posterior standard deviation, in terms of θ, of parameter i.

A *prima facie* model selection criterion for models that are capable of representing the data (those having good values of χ^2/m) is to select the model with the smallest value of U. This might seem at odds with choosing the model with the largest evidence because of the seeming correspondence of U and $V^{1/n}$. However, one needs to remember that, for the same maximum likelihood, because of parameter correlation, $V^{1/n}$ need not track U. For the 2-parameter example of Exercise 1, $V^{1/2} = U(2\pi)^{1/2}(1-\gamma^2)^{1/4}$, where γ is the correlation between parameters 1 and 2. Also if V is sufficiently small (it is always less than 1), the evidence goes down with n (simpler model preferred) even if $V^{1/n}$ increases with n.

Methods for evaluation of the evidence integral. Returning to Eq. (3) itself, it is not obvious how to evaluate the required integrals in high dimension situations, where direct numerical integration is not practical. This is the subject considered in the remainder of this Chapter.

We consider two methods for evaluation of the evidence integral.

The first method considered (Miller et al. 2007) is to add another variable that determines the hypothesis: the normal one or a reference hypothesis in which $L(\theta)$ has a constant reference value L_{ref}. Then, using MCMC, one calculates the fraction of the time the chain is in the reference hypothesis rather than the normal hypothesis, and

Method I: $E = L_{ref} \dfrac{Avg(I_{hyp})}{1 - Avg(I_{hyp})}$,

where the function $I_{hyp}(\theta)$ takes the value 1 when the chain is in the normal hypothesis and 0 when the chain is in the reference hypothesis.

The difficulty with this method comes when the energy of the reference hypothesis is large (L_{ref} small), so that the normal chain rarely has an energy of this magnitude. This causes slow mixing from normal to reference hypothesis, because the current point of the chain must be far out in the tail of the distribution function to make the transition from the normal to the reference hypothesis, and the current point is unlikely to visit this tail region. The reverse transition from the reference hypothesis is also unlikely because it requires a point from the candidate distribution to land fairly near the mode of the normal hypothesis, and for transitions from the reference hypothesis, the candidate distribution extends over the entire space of the normal hypothesis. See also the subsequent discussion of "Temperature Methods".

156

To motivate the second method, consider the basic MCMC average defined as

$$Avg(f) \equiv \frac{\int d\theta \, L(\theta) f(\theta)}{\int d\theta \, L(\theta)}$$

with $f(\theta) = 1/L(\theta)$, which gives

$$E = \int d\theta \, L(\theta) = \frac{\int d\theta \, L(\theta)}{\int d\theta \, L(\theta)/L(\theta)} = \frac{1}{Avg(1/L)} \quad .$$

This approach is discussed in a review of Bayesian hypothesis testing given in Gilks et al., Chapter 10 (Gilks et al. 1996). The average is evaluated using MCMC in the usual way.

Clearly, when the support region of the likelihood function is a small fraction of the entire parameter space, as is the case for the models considered in this book, this method will not work.

Our second method is similar to this method in that an independent evaluation of an MCMC-average numerator is used to calculate the MCMC-average denominator. This method starts with the equation

Method II: $E = \int d\theta \, L(\theta) = \dfrac{\int d\theta \, L(\theta) I_{l_c}(\theta)}{Avg(I_{l_c})}$,   (5)

where the function $I_{l_c}(\theta)$ takes the value 1 when $L(\theta) > L_c$ and 0 otherwise, with L_c some critical value of the likelihood function close to the maximum value.

The denominator in Eq. (5) is calculated as an MCMC chain average using the tape file in the usual way. Monte Carlo integration is used to calculate the numerator in Eq. (5); however, first the smallest possible region of integration is determined. This is the smallest multidimensional box that just contains the support region of $I_{l_c}(\theta)$. Integration over just this box does not change the value of the integral, because the integrand is zero outside. The integration box is found by examination of the tape file after burn in, finding the minimum and maximum for each parameter θ_i for all tape-file records $k = 1 \ldots N$ with $L(\theta_i^{(k)}) > L_c$. The volume of the integration box is then

$$V_{box} = \prod_{i=1}^{n} \left(\max_{k=1}^{N} \left(\theta_i^{(k)} \right) - \min_{k=1}^{N} \left(\theta_i^{(k)} \right) \right) \ .$$

Monte Carlo integration is used to evaluate the integral, using samples $\theta^{(l)}$ generated randomly inside the box. Choosing the number of Monte Carlo integration points as N_{MC}, the result is

$$\int_{box} d\theta \, L(\theta) I_{L_c}(\theta) \cong V_{box} \frac{\displaystyle\sum_{l=1}^{N_{MC}} L(\theta^{(l)}) I_{L_c}(\theta^{(l)})}{N_{MC}} \ .$$

Therefore the desired Evidence integral is given by

$$E = \frac{\displaystyle\int_{box} d\theta \, L(\theta) I_{L_c}(\theta)}{Avg(I_{L_c})} = V_{box} \frac{N_0}{N} \frac{\displaystyle\sum_{l=1}^{N_{MC}} L(\theta^{(l)}) I_{L_c}(\theta^{(l)})}{N_{MC}} \ ,$$

where N_0 is the total number of burned-in records in the tape file.

Having a large number of model parameters necessitates a very large number N_{MC} of Monte Carlo trials to get some reaonable minimum number (say 10) of nonzero terms in the summation, because, for high-dimension situations, the volume of the inscribed ellipsoid or sphere is a small fraction of the volume of the box (see Exercise 3).

Figure 1 shows results for the linear, 2-parameter problem discussed in Exercise 1. Note the small region of θ space shown, implying that the region inside the box is quite small relative to the entire prior domain.

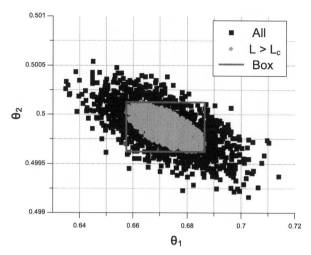

Figure 1—Scatter plot for example problem showing all points from the posterior distribution on the tape file, those with $L(\theta) > L_c$, and the boundaries of the box.

A similar plot for the 4-parameter nonlinear problem discussed subsequently (Test Problem 2 in the Section "Tests of the methods for continuous variables") is shown in Fig. 2

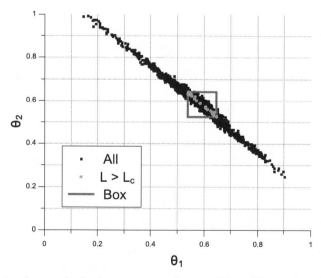

Figure 2—Scatter plot for 4-parameter nonlinear problem.. The number of Monte Carlo integration points generated uniformly inside the box needed to produce 10 nonzero terms in the summation was about 20000, showing the small fraction of the box occupied by the inscribed ellipsold.

Figure 2 illustrates that a large number of Monte Carlo trials may be required for the integration, and it also illustrates why importance sampling using samples from the posterior would not work for this problem (see Exercise 2). In importance sampling the integration makes use of the forward model points already calculated and recorded in the tape file. Importance sampling would assign integration weight to the points from the tape file equal to the inverse of the likelihood function, which requires that the likelihood function not be too small for any point. As seen in the figure, large regions of the box are clear of points from the tape file, which means that in these regions the likelihood function is very small.

As can be appreciated from the example shown in Fig. 2, this method of evaluating the evidence integral can be difficult or impossible for high-dimensional problems.

So, both methods I and II of evaluating the evidence integral considered here do not match the power of the basic MCMC calculation. However, the uncertainty calculation using Eq. (4) does. It involves straightforward and fast additional processing using the tape file.

Analysis of the reference hypothesis solution for finite integer-valued Markov Chains. To understand more clearly the use of a reference hypothesis, let us first consider a discrete Markov Chain, as was done in Chapter 9. The eigenvalues and eigenvectors of the transition matrix are calculated, and the time to reach steady state is obtained from the magnitude of the second largest eigenvalue. As discussed in Chapter 9, the initial state can be represented as a linear combination of eigenvectors, and each eigenvector has time dependence λ^t, where λ is the eigenvalue and the t is the chain iteration number. The largest eigenvalue is always 1, and this eigenvalue-1 eigenvector gives the steady state. The second largest eigenvalue is the longest persisting of all the others. It has time dependence $\lambda^t \propto \exp(-t/\tau)$, where $\tau = -1/\log(\lambda)$.

The first approach to the task of including a reference hypothesis is to expand the parameter space from $i = 1,... n$ to $i = 0,... n$, where $P_0 = P_{ref}$ is the probability of being in the reference hypothesis. We will reorder the indices as $i = 1,...n, 0$ with 0 last.

We define a new transition matrix given in block form by

$$
T = \begin{bmatrix} T_{i,j} & T_{i,0} \\ T_{0,j} & * \end{bmatrix}
$$

where $T_{i,j}$ is the transition matrix for the normal hypothesis as discussed in Chapter 10, however with the off-diagonal terms multiplied by a constant $a = 1 - q_{switch}$, where q_{switch} is the candidate probability of switching from the normal hypothesis to the reference hypothesis (the diagonal terms of $T_{i,j}$ are determined by column normalization as usual).

Similarly there is a candidate probability of switching from the reference hypothesis to the normal hypothesis, but this is taken to be 1 to minimize the time pointlessly spent in the reference hypothesis. The symbol $*$ denotes the scalar needed for normalization of the last column, given by 1 minus the sum of the off-diagonal elements.

The candidate distribution for transitions $0 \rightarrow i$ is a constant for all values of i,

$$
q_{i,0} = \frac{1}{n} \quad ,
$$

which sums to 1 over the n states i. For transitions $j \rightarrow 0$,

$$
q_{0,j} = q_{switch} \quad .
$$

The acceptance probabilities for the MRT algorithm are given from Eq. (16) of Chapter 9 by

$$\alpha_{i,0} = \min\left(1, \frac{n\overline{P}_i q_{switch}}{P_0}\right)$$

$$\alpha_{0,j} = \min\left(1, \frac{P_0}{n\overline{P}_j q_{switch}}\right)$$

(7)

Thus

$$T_{i,0} = \alpha_{i,0} q_{i,0} = \frac{1}{n} \min\left(1, \frac{n\overline{P}_i q_{switch}}{P_0}\right)$$

$$T_{0,j} = \alpha_{0,j} q_{0,j} = q_{switch} \min\left(1, \frac{P_0}{n\overline{P}_j q_{switch}}\right)$$

The MC algorithm is simpler, does not use the hypothesis switching candidate distribution, and

$$T_{i,j} \propto \overline{P}_i \quad,$$

now including $i = 0$, with column-by-column normalization,

$$T_{i,j} = \frac{\overline{P}_i}{\sum_{i'=0}^{n} \overline{P}_{i'}} \quad,$$

for all the elements of the transition array within the same template of nonzero elements as used for the MRT algorithm.

Figure 3 below shows the eigenvalue-1 eigenvector with $\overline{P}_0 = 0.1$, using the MC algorithm. The full width of the random walk for the normal hypothesis $(l = 20)$ is 5 times the standard deviation of the assumed steady-state distribution $(\sigma = 4)$.

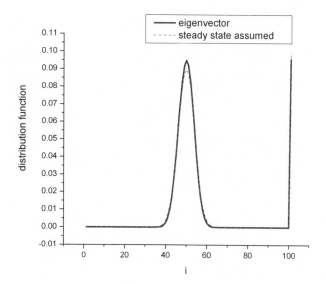

Figure 3—Eigenvalue-1 eigenvector, using the MC algorithm.

For the MRT algorithm there is very good agreement between the eigenvalue-1 eigenvector and the assumed steady state, as must be true.

Table 2 shows the times to reach a steady state, calculated from the second largest eigenvalue for the MRT and MC algorithms. As in Fig. 3, the diagonal band width l is 20 and the standard distribution of the steady-state distribution is 4.

Table 2—Time to reach steady state (3τ) for the MRT and MC algorithms with a reference hypothesis.

algorithm	3τ
MRT	13.04
MC	1.63

Temperature methods. An alternative to the single-point reference hypothesis approach is to replicate the entire parameter space $i = 1,... n$ into $i = n+1, 2n$, where these new points have the constant energy of the reference hypothesis. This situation can be understood as similar to a temperature algorithm (with energy divided by temperature-replacing energy). Temperature algorithms are used to facilitate mixing between isolated modal regions of the distribution function by allowing transitions to and from a higher temperature replica of the system where the modal regions are more connected. Introduction of a reference hypothesis has the same effect in facilitating transitions between isolated modal regions, and in this sense is similar to a temperature algorithm with only 2 temperatures. Regarding a reference hypothesis with constant energy, discrete chain calculations do not indicate an advantage in replicating the entire space over having a single-point reference hypothesis.

For a two-temperature algorithm, discrete chain calculations show that the temperature switching probabilities can be small without having a large adverse effect on the time to reach steady state. Thus one can imagine running chains with different temperatures separately and independently and then, at some intervals, switching temperatures. An interesting approach is to replicate the initial space a large number of times with a sequence of temperatures (Hukushima and Nemoto 1996) using parallel calculations for the different temperatures, and then, at some interval, switching the temperatures. Calculations with a two-temperature algorithm show that when the energy of the system is large, unless the temperature difference is small, transitions need to occur through a region of exponentially small values of the distribution function $\exp(-E/T)$ in moving from temperature to temperature. If the energy is large, a sequence of temperatures spaced closely enough to allow good mixing is needed.

Temperature algorithms will not be discussed further in this book.

Continuous variables. Going from a discrete Markov Chain to a Markov Chain with continuous variables requires the correspondence

$$\overline{P}_i \leftrightarrow L(\theta_i)\Delta\theta = L(\theta_i)/n \quad ,$$

where n is the number of discrete points, so that $n\overline{P}_i \leftrightarrow L(\theta_i)$.

Also, to apply the usual formulas, the energy in the reference hypothesis needs to be modified in this way (see Exercise 5).

$$E_{ref} \rightarrow E_{ref} + \log(q_{switch}) \quad .$$

In the continuous version of the discrete MRT algorithm with a reference hypothesis, transitions from the reference hypothesis occur to the entire space of the normal hypothesis. For these transitions all parameters must be moved together (no grouping), and the random-walk region is the entire space.

The continuous version of the discrete MC algorithm with a reference hypothesis usually doesn't work very well, even though the discrete version does work well. In the discrete MC algorithm, in the denominator of the transition matrix sums over P_i appear. In the continuous version these discrete sums should approximate integrals. The difficulty comes in the correspondence between the discrete sum and the integral

$$\sum_i P_i \rightarrow \frac{2\Delta}{l} \sum_i L(\theta_i) \cong \int d\theta L(\theta) \quad .$$

If Δ is too large compared to the support region of L, the numerical approximation of the integral by the discrete sum is poor. This happens for the transitions from the reference hypothesis to the normal hypothesis where $\Delta = 1$, which is too large in most cases. This kind of thinking leads to another method, discussed in Ref. 4 but not considered here, where one directly uses the numerical approximation of the integral by the discrete sum when Δ is large enough to encompass the support region of P.

Tests of the methods for continuous variables. Two test problems are considered, both involving simulated data that is a function of time. For the examples discussed here the number of data points was 10 and the data variation was normal with specified ratio of average data standard deviation to true data vaule = 0.1.

In the first problem (see Exercise 1), the model funtion is a polynomial in time and the model parameters are the polynomial coefficients. This is a linear model, because the model function is a linear function of the parameters, and the prior is linear-uniform, so that the model function is a linear function of the θ parameters. For this problem, the data are generated assuming only one term in the polynomial, that is, the true value of the data is a constant. Figure 4 shows data versus time together with the MCMC interpretation, assuming 2 model parameters (constant and slope). The MCMC interpretation yields the posterior mean and posterior standard deviation shown in the plot.

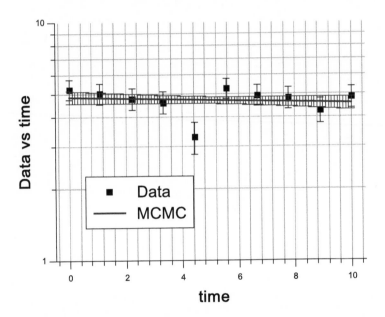

Figure 4—Simulated data and MCMC interpretation. The true value of the data is constant and the model has two parameters: a constant value and a linear slope. The posterior mean and posterior standard deviation of the MCMC interpretation are shown.

In the second example problem, the model function is a sum of decreasing exponentials. The coefficients in the sum have linear-uniform priors and the rates have log-scale uniform priors. The data are generated assuming only one exponential term in the sum. Figure 5 shows data versus time together with the MCMC interpretation, assuming 2 exponential terms (4 model parameters).

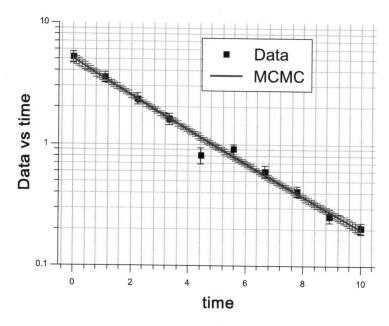

Figure 5—Simulated data and MCMC interpretation. The true values of the data are from a single decreasing-in-time exponential. The model has 4 parameters: coefficents and rates for 2 exponential terms. The posterior mean and posterior standard deviation from the MCMC interpretation are shown.

A summary of results is shown in Table 3. These calculations were done using the program GMCMC included in the supplementary material for this Chapter.

Table 3—Summary of calculation results for the two example problems. The number of model parameters is n, and the number of data points is $m = 10$. The chain average of χ^2 / m is shown together with the uncertainty U and three calculations of the evidence integral: using the methods I and II described in this chapter and using direct, nested, 1-dimensional numerical integration.

Problem	n	$\langle \chi^2 / m \rangle$	U	$\log(E)_I$	$\log(E)_{II}$	$\log(E)$
1	1	1.31	0.0054	-10.3	-10.7	-10.3
	2	1.36	0.0041	-15.7	-16.2	-15.6
	3	1.33	0.0036	-21.6	-21.8	-21.5
2	2	1.36	0.012	-13.3	-13.1	-13.4
	4	1.33	0.14	-14.4	-14.2	-14.3
	6	1.32	0.21	-15.1	-15.7	

As can be seen from Table 3, the model with the largest evidence is the correct one. For Problem 2, the minimum uncertainty model clearly selects the correct model. For Problem 1, this is not the case. However for this linear problem, the redundant coefficients are zero within uncertainties, and would naturally be eliminated on this basis.

Exercises

1. Assume a forward model as a function of time that is constant, $f(t) = a$ (hypotheses I) or has linear time dependence $f(t) = a + bt$ (hypotheses II). The data are from some unknown function of time plus Gaussian measurement uncertainties (standard deviation σ_j for data point j). Assuming uniform priors with full extents Δ_a and Δ_b, show that the evidence integrals are given by

$$E(I) = L_{\max} \sqrt{2\pi \langle \sigma^2 \rangle / m} / \Delta_a$$

$$E(II) = L_{\max} \left(2\pi \langle \sigma^2 \rangle / m \right) / \left(\Delta_a \Delta_b \sqrt{\langle t^2 \rangle - \langle t \rangle^2} \right)$$

where m is the number of data points. The average denoted by angle brackets is defined by

$$\langle g(t) \rangle = \sum_{j=1}^{m} \frac{g(t_j)}{\sigma_j^2} / \sum_{j=1}^{m} \frac{1}{\sigma_j^2} \quad,$$

for an arbitrary function of time $g(t)$.

2. Show how the evidence integral might be calculated using importance sampling.

3. For an approximately linear model $L(\theta) \cong L_{max} \exp(-\tilde{\theta}^T H \tilde{\theta} / 2)$, where $\tilde{\theta}$ is the deviation from the maximum, show that
$E = L_{max} (2\pi)^{n/2} / \sqrt{\det(H)}$.

4. Show that the volume V of an n-dimensional sphere of radius R is given by $V = \pi^{n/2} R^n / (n/2)!$, and therefore the fraction of an n-dimensional box occupied by an inscribed sphere becomes very small as the dimensionality is increased.

5. Using Eq. (7), show that in the continuous case the probability for accepting transitions that switch hypothesis is given by the usual formulas except that the energy in the reference hypothesis is changed in this way: $E_{ref} \rightarrow E_{ref} + \log(q_{switch})$, where q_{switch} is the candidate probability of switching from the normal hypothesis to the reference hypothesis.

References

1. Miller, G., H. Martz, T. Little, and L. Bertelli, "Bayesian Hypothesis Testing--Use in Interpretation of Measurements", Health Phys. 94(3):248-254 (2008).

2. Gilks, W. R., S. Richardson, and D. J. Spiegelhalter. *Markov Chain Monte Carlo in Practice*. New York: Chapman and Hall, 1996.

3. Hukushima and Nemoto. "Exchange Monte Carlo Method and Application to Spin Glass Simulations." *J. Phys. Soc. Japan*. 65:1604 (1996).

4. Miller, G. "Multidimensional Integration of a Positive Function Using Markov Chain Monte Carlo." *The Open Numerical Methods Journal* 3:20–25 (2011).

Chapter 15. Example—Detection of a Rarely Occurring Contaminant

This is a situation that is quite common: measurements are done to detect a contaminant or pathogen of concern. The occurrences of the contaminant are thankfully rare, and the civil-life client wants to know if *any* of the material is present. And, there is a background level. This may be an apparent background level of the material caused by the measurement process, or an actual background level. An example of the latter would be contamination of a watershed by some upstream source. The actual background in that case would be the naturally occurring level of the contaminant present in all watersheds in the area. So the question becomes, "Is the contaminant present above background?" In this way we are able to attribute the contamination to the source of interest. The example discussed in this chapter is from internal dosimetry, but the basic ideas apply in any situation of this type.

For this example, a worker is being routinely monitored for intakes of a radionuclide, which is a material encountered in the workplace. The workplace is engineered to minimize the chances of intakes, and no level of intake is considered acceptable. Bioassay samples (for example, urine or fecal samples) are taken and analyzed for the presence of the nuclide. What then? How does one model or interpret the measurement?

The prior. There is a time interval Δt when the intakes might have occurred. We assume that the distribution of the total amount of the intake, I, is given by the alpha distribution,

$$P(I)dI = p\left(\frac{I}{I_{max}}\right)^p \frac{dI}{I} \, , \tag{1}$$

where $p \equiv \alpha\Delta t$ is small and represents the probability of intakes, and I_{max} is the maximum conceivable amount of intake. The experiment underlying this distribution is to select a workplace and worker from a collection of "clean" workplaces and "normal" workers and workplace situations and to measure the intake. It is known from experience that significant intakes are extremely rare. Equation (1), while not the actual result of a real experiment, is used to summarize real experimental experience. Mathematically, Eq. (1) has a delta-function-like character with almost all the probability concentrated at 0, but away from 0 having an approximately log-space uniform distribution,

$$P(I)dI \cong p\, d\log(I), \quad I > 0 \quad . \tag{2}$$

This prior is an example of a rather extreme prior that can significantly influence the interpretation of the measurement, hence the importance of the word "rarely" in the title of this chapter.

The forward model, from intake to measurement quantity, which is assumed to be linear with intake amount, in this case is just a numerical coefficient denoted by F_l where the index l distinguishes different types of materials, which have different biokinetics, and hence for the same intake, cause different amounts of the measured bioassay quantity. By letting $I \rightarrow F_l I = \psi$, Eq. (1) can also be used to describe the probability distribution of the true value of the measured quantity, which we denote by ψ.

The posterior. The measurement is assumed to be a counting measurement with N counts detected in sample counting period T. Background measurements are discussed in Chapter 4. The background prior or the "moving-target" characterization of the background is assumed to be a gamma distribution with parameters α_B and β_B. For a background measurement of N_B counts in background counting period T_B interpreted using a flat prior, $\alpha_B = N_B + 1$ and $\beta_B = T_B$. The background scaling factor is denoted by R,

$$R = \beta_B / T \quad .$$

Using the exact likelihood function results of Chapter 7, the posterior distribution of ψ given the data is the product of the prior times the likelihood

$$P(\psi \mid data)d\psi \propto p\left(\frac{\psi}{\psi_{max}}\right)^p d\log\psi \times L(\psi)$$

$$L(\psi) \propto \int d\log A \exp\left(-\frac{1}{2}\left(\frac{\log A/A_0}{S_{norm}}\right)^2\right) \qquad (3)$$

$$\times e^{-A\psi} \sum_{k=0}^{N} \frac{N!(N-k+\alpha_B-1)!}{k!(N-k)!}\left(A\psi(1+R)\right)^k$$

Recognizing that Eq. (3) involves gamma distributions, we rewrite it in terms of normalized gamma distributions using

$$\psi^{k-1+p}e^{-A\psi} = \frac{(k+p-1)!}{A^{k+p}}g(\psi \mid k+p, A) \quad ,$$

which gives

$$P(\psi \mid data) \propto \int \frac{d\log A}{(A\psi_{max})^p}\exp\left(-\frac{1}{2}\left(\frac{\log A/A_0}{S_{norm}}\right)^2\right)$$

$$\times pN!\sum_{k=0}^{N}\frac{(N-k+\alpha_B-1)!(k+p-1)!}{k!(N-k)!}(1+R)^k g(\psi \mid k+p, A)$$

Notice that in the limit $p \to 0$ only the $k=0$ term survives in the summation by virtue of the fact that $p(k+p-1)! = p! \to 1$. Because $(A\psi_{max})^p \cong 1$ for small p, for the $k=0$ term the $d\log A$ integral is approximately the lognormal normalization integral, and this term is $\sqrt{2\pi}S_{norm}(N+\alpha_B-1)!$ times the gamma distribution

$g(\psi \mid p, A) \cong \delta(\psi)$, where $\delta(\psi)$ is the delta function at 0. Making this replacement we obtain

$$P(\psi \mid data) \propto \delta(\psi) + \int \frac{d \log A}{\sqrt{2\pi} S_{norm}} \exp\left(-\frac{1}{2}\left(\frac{\log A / A_0}{S_{norm}}\right)^2\right)$$

$$\times p\left\{\frac{N!}{(N + \alpha_B - 1)!} \sum_{k=1}^{N} \frac{(N - k + \alpha_B - 1)!}{k(N - k)!}(1 + R)^k g(\psi \mid k, A)\right\}$$

$$\tag{4}$$

There are several parameters entering into Eq. (4), which cause its complex appearance, but one sees that it is a delta function at 0 coming from the prior, plus distributions peaked at the "measured values" corresponding to the measured counts (see Exercise 1).

One can appreciate the important role of the prior, which in the limit $p \to 0$ pulls the posterior toward a concentration at 0 versus the number of detected counts N, which pulls the posterior up toward the nominal measurement value.

Is the contaminant present? One way to address the qualitative question of interest, "Is the contaminant present?" is by a hypotheses test. Hypothesis 1 is what we have so far discussed, where there is some nonzero intake. Another hypothesis 0 is that there is no amount of the contaminant above background, that is, $\psi = 0$. Integrating Eq. (4) with respect to $d\psi$, where the normalized gamma distributions are replaced by 1, gives us the answer for the relative probabilities of hypothesis 1 and hypothesis 0:

$$\begin{aligned} P(Hyp1 \mid data) &\propto 1 + pC(N, \alpha_B, R) \\ P(Hyp0 \mid data) &\propto 1 \end{aligned} \tag{5}$$

where we have defined a count quantity

$$C(N, \alpha_B, R) = \frac{N!}{(N + \alpha_B - 1)!} \sum_{k=1}^{N} \frac{(N - k + \alpha_B - 1)!}{k(N - k)!}(1 + R)^k . \tag{6}$$

Equal prior probabilities of the two hypotheses are assumed. A 50% posterior probability for hypothesis 0 means that the two hypotheses are equally likely, given the data, and are therefore indistinguishable. Notice that for equal prior probabilities the posterior probability of hypothesis 1 in the limit $p \to 0$ equals the posterior probability of hypothesis 0. In order to imply a larger probability for hypothesis 1, the count quantity $C(N, \alpha_B, R)$ must be large, overcoming the smallness of p.

It is interesting that there is no dependence on S_{norm} in Eq. (6).

Using the normalization condition

$$P(Hyp1 \mid data) + P(Hyp0 \mid data) = 1 \quad,$$

we obtain

$$P(Hyp0 \mid data) = \frac{1}{2 + p\,C(N, \alpha_B, R)} \quad, \qquad (7)$$

and for the odds ratio for hypothesis 1,

$$O = \frac{P(Hyp1 \mid data)}{P(Hyp0 \mid data)} = 1 + pC(N, \alpha_B, R) \quad.$$

This then is a scientifically responsible way to provide what the civil-life client is seeking. It is obvious to the client that the entire point of the measurement is to tell him or her whether the contaminant is present. Studies have shown that the client, while not pretending to understand the technical jargon he or she is sometimes given, of course believes that his or her "bottom line" question has been addressed and will find a way to interpret the jargon accordingly.

The priors are provided by "informed guesses" if necessary. The scientist understands the role the prior plays in the interpretation of the measurement. If the influence of the prior is significant and the prior is not well characterized by other measurements, the scientist insists on making more measurements until the influence of the prior is not incommensurate with actual knowledge of the prior. To simulate more measurements using Eq. (6), we can increase the count time T. If this is done, the number of counts N and α_B (N_B) will increase and eventually cause the truth to be revealed.

These calculations can also be carried out using MCMC, and the comparison is instructive. In MCMC one uses uniformly distributed "theta" variables between 0 and 1 that incorporate their respective priors. The "theta" variable describing the intake or true value is given by

$$\theta = \left(\frac{\psi}{\psi_{max}}\right)^{p} \quad .$$

$$\psi = \psi_{max}\,\theta^{1/p}$$

(8)

The extreme nature of this function can be seen by noticing that the power in Eq. (8) might be on the order of 1000. This means that only values of θ very close to 1 contribute reasonable nonzero values of ψ. The random-walk step Δ therefore must be quite small to effectively explore this region. However, the chain must also explore the region corresponding to $\psi = 0$, and the chain must go back and forth between these two regions many times. A way to handle this situation, which works well in practice, is to probabilistically choose either a random walk with small step Δ or exploration of the entire space ($\Delta = 1$). This is the mixed Δ method that was discussed in Chapter 11.

Hypothesis testing can be done in the MCMC calculation (Method I of Chapter 14), with two hypotheses considered. In the first, hypothesis 0, no intakes were assumed to have occurred, and the true value of the bioassay quantity is zero. This hypothesis has no parameters, and the likelihood function for $\psi = 0$ (the "flat line" interpretation of the data) can be calculated once and for all at the beginning of the calculation. In the second hypothesis, hypothesis 1, an intake was assumed in the time interval Δt preceding the bioassay measurement. The posterior probabilities of the two hypotheses are given by their fractions of chain time.

In Table 1 we show the probability of hypothesis 1, P, the odds for hypothesis 1, and the odds $O = P/(1-P)$ minus 1, calculated using MCMC for three values of p. This is for $N = 7, \alpha_B = 3, R = 4.3$ for which the count quantity C from Eq. (6) is equal to 1000. The probability of hypothesis 1 is the fraction of the time the chain spends in hypothesis 1.

Table 1—Independent calculation of the odds ratio for hypothesis 1 using MCMC. The check is that the odds ratio minus 1 equals 1000 times p.

p	$P(Hyp1)$	Odds-1	$1000p$
0.001	0.670 ± 0.027	1.03 ± 0.25	1
0.003	0.794 ± 0.0024	2.85 ± 0.06	3
0.01	0.9144 ± 0.0063	9.69 ± 0.86	10

The MCMC calculations obtained the likelihood function from an interpolation table of $\chi(\psi)$ constructed from the exact likelihood function values calculated using numerical integration (the default) or Monte Carlo integration, so this is an independent check of the analytical results.

In Fig. 1 we show the cumulative probabilities of true amount in the sample in these three cases. These distributions were obtained from a 1000-record "tape" file.

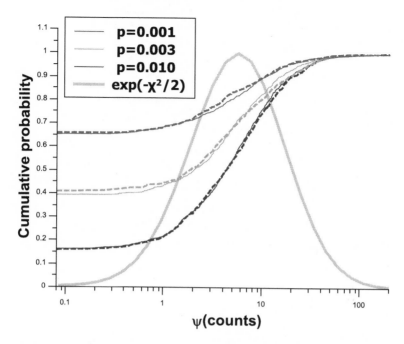

Figure 1—Cumulative probability of the true amount in the sample for three values of the prior parameter p , calculated using MCMC with hypothesis testing. There are two runs to demonstrate convergence in each case. The lognormal normalization uncertainty is $S_{norm} = 1$ (GSD=2.7). Also shown is the likelihood function $\exp(-\chi^2 / 2)$.

The posterior probability of zero true amount is given by

$$P(\psi = 0 \mid data) = P(\psi = 0 \mid Hyp0, data)P(Hyp0 \mid data)$$
$$+ P(\psi = 0 \mid Hyp1, data)P(Hyp1 \mid data) \quad .$$
$$= \frac{2}{2 + pC(N, \alpha_B, R)}$$

(9)

For the same numbers as in Fig. 1 and the count quantity $C = 1000$, for $p = 0.001, 0.003$ and 0.01, Eq. (9) and the left most points on the curves in Fig. 1 should equal 2/3, 2/5, and 1/6, which is seen to be the case.

Alternative method, not using hypothesis test. For only a single hypothesis,

$$P(\psi = 0 \mid data) = \frac{1}{1 + pC(N,\alpha_B,R)} \qquad , \qquad (10)$$

which follows directly from Eq. (4). Equation (10) offers an alternative to the probability of hypothesis 0 as a definition of "nothing detected." This concept is easily grasped visually from a graph such as Fig. 1 and is perhaps clearer than hypothesis testing in that it provides a number near 1 when nothing is detected rather than 0.5 (meaning hypothesis 0 is indistinguishable from hypothesis 1). Also, using this approach eliminates the need for hypotheses testing in the MCMC code.

These probabilities have the usual meaning. If the experiment were to be repeated a large number of times with samples drawn from the prior probability distribution, the posterior probability is that fraction of times where the stated outcome occurs, given the measurement results.

Exercises

1. Derive a general expression for the posterior probability of true amount ψ for a counting measurement with a gamma distribution prior. What is the limit that gives the situation discussed in this chapter?

2. According to Ref. 1, *"The decision threshold,* y^**, of the non-negative measurand…quantifying the physical effect of interest, is the value…which allows the conclusion that the physical effect is present, if the primary measurement result,* y*, exceeds the decision threshold,* y^**. If the result,* y*, is below the decision threshold,* y^**, the result cannot be attributed to the physical effect, nevertheless it cannot be concluded that it is absent. If the physical effect is really absent, the probability of taking the wrong decision, that the effect is present, is equal to the specified probability,* α *…"*. For $\alpha = 0.05$, following the guidance of Ref. 1, the decision threshold is 1.645 times the standard deviation for zero true result. Show that in terms of net counts
    $y = N - N_B / R$, $\sigma = \sqrt{N + N_B / R^2}$ this is
    $y^* = 1.645\sqrt{N_B(R+1)}/R$, where N_B is the number of measured

background counts and R is the ratio of background count time to sample count time.

3. Assume $N_B = 16$, $R = 5$, and sample counts $N = 10$. Show that $y / y^* = 2.11$ so that this sample "allows the conclusion that the physical effect is present" with a "probability of taking the wrong decision" of 0.05.

4. Using a spreadsheet, calculate the probability of the sample being zero using Eq. 9 assuming the prior parameter $p = 0.001$ and show that it is 96%. One sees that the interpretation of this measurement based on the existing international standard is 1) rather confusing regarding probabilities (the probability statement is predicated on "*If the physical effect is really absent,*" which is unknown) and 2) is not consistent with the probabilistic interpretation, which depends on the prior.

5. Using the formula for y and σ from Exercise 2, solve for N given y / σ, N_B, and R. Now using same spreadsheet calculation as in the previous exercise, for $y / \sigma = 4$ and $p = 0.001$ show that the posterior probabilities of the sample being zero for $N_B = 50$, $R = 10$ and $N_B = 1$, $R = 0.1$ are 0.01% and 98%, which shows the inadequacy of recording data as result and standard deviation only.

6. In Ref. 1 a confidence interval is defined as "*The limits of the confidence interval are provided for a physical effect…in such a way that the confidence interval contains the true value of the measurand with the specified probability…*" Comment on this construct in light of the fact that it does not involve a prior probability distribution.

7. What are the left intercept values corresponding to Fig. 1 if there is only hypothesis 1? Check by running the Fortran program ID supplied with the supplementary material for this chapter.

8. Using the uranium method blank data shown in Fig. 4 of Chapter 4 and included in the supplementary material for Chapter 4, what is the number of uranium sample counts needed for 95% detection probability using the "fixed target" and the "moving-target" characterizations of the background?

Reference

1. INTERNATIONAL STANDARD ISO 11929, "Determination of the characteristic limits (decision threshold, detection limit and limits of the confidence interval) for measurements of ionizing radiation— Fundamentals and application." First edition, 2010-03-01.

Chapter 16. Example—Modeling the Intake of a Radionuclide from the Goiânia Accident

In September 1987, in Goiânia, a city in the central part of Brazil, a radiation therapy unit from an abandoned medical facility was scavenged by a few individuals looking for something of economic value, and the source capsule containing 50.9 TBq of ^{137}Cs was broken apart (a Bq is one radioactive disintegration per second, T means 10^{12}). Intrigued by the luminescent qualities of the powder they found, and not recognizing any danger, they spread the contamination widely for about a week before the first medical symptoms appeared, and the cause was discovered. About 110,000 persons were monitored during the first screening using a Geiger-Müller detector, and 249 of them were identified as internally or externally contaminated. They were immediately submitted to external decontamination treatment. The highest radiation doses result from internal contamination, that is, intakes where the radionuclide was taken into the body. The chemical form of ^{137}Cs source was cesium chloride, which is very rapidly absorbed into the bloodstream whether the route of intake is ingestion or inhalation.

Forward model. The biokinetic model for cesium recommended at that time by the International Commission on Radiation Protection (ICRP) is shown in Fig. 1.

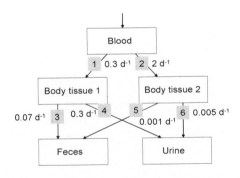

Figure 1—Biokinetic model for cesium recommended by the International Commission on Radiation Protection. The nominal transfer rates from compartment to compartment are indicated.

180

The model shown assumes the intake is transferred immediately to the blood and, for simplicity, eliminates the urinary bladder and gastro-intestinal tract. The radiation dose is caused by the radioactive decay of the cesium (30-year half-life), which results in a fairly uniform irradiation of the whole body. The radiation dose is proportional to the number of radionuclide disintegrations occurring in the body.

In this case the forward model is simple enough so that it can be solved analytically. The model shown schematically in Fig. 1 consists of 3 linear equations in 3 unknowns: the content in Blood (B) and the contents in Body Tissues 1 and 2 (S and T). These three unknowns are represented by a column vector ψ:

$$\psi = \begin{bmatrix} B \\ S \\ T \end{bmatrix} .$$

The forward model equations are of the form

$$\frac{d\psi}{dt} = A\psi$$

$$A = \begin{bmatrix} a & 0 & 0 \\ d & b & 0 \\ e & 0 & c \end{bmatrix}$$

where

$$a = -(k_1 + k_2 + \lambda)$$
$$b = -(k_3 + k_4 + \lambda)$$
$$c = -(k_5 + k_6 + \lambda)$$
$$d = k_1$$
$$e = k_2$$

with the transfer rates k as indicated in Fig. 1 and λ the radioactive decay rate. The solution of this system of linear differential equations is a sum of 3 terms proportional to $e^{-\gamma_i t}$ where γ_i is the ith eigenvalue of the matrix A. The eigenvalues are the 3 solutions of the 3rd order polynomial equation

$$\det(A - \gamma I) = \det \begin{bmatrix} a - \gamma & 0 & 0 \\ d & b - \gamma & 0 \\ e & 0 & c - \gamma \end{bmatrix} = 0 \ ,$$

which one sees are $\gamma = a, b,$ and c. The eigenvectors are also easy to determine.

The initial conditions, for unit intake, are that there is an initial amount 1 in B and nothing initially in S and T. Therefore the solution of the forward model for unit intake is given by

$$\psi(t) = \begin{bmatrix} 1 \\ d/(a-b) \\ e/(a-c) \end{bmatrix} e^{at} + \begin{bmatrix} 0 \\ -d/(a-b) \\ 0 \end{bmatrix} e^{bt} + \begin{bmatrix} 0 \\ 0 \\ -e/(a-c) \end{bmatrix} e^{ct} \ .$$

The model calculation of the bioassay data (U is 24-hour urinary excretion; W is whole body retention) is given by

$$U = I(k_4 S + k_6 T)$$
$$W = I(B + S + T)$$

where I is the intake in activity units, the same units as the bioassay data. Activity is the number of atoms times the radioactive decay rate λ in Bq or sec^{-1}, and it gives the number of nuclear transformations per sec.

182

The parameters of the forward model calculation are 7 in number, consisting of the 6 rate coefficients shown in Fig. 1 and the amount of intake I. The problem is to determine these parameters, given the bioassay data, but the bottom-line quantity of interest is the radiation dose, which is proportional to the total number of nuclear transformations that will occur in the whole body,

$$N_t = \int_0^\infty dt\, W(t)$$

$$N_t = -C\frac{I}{a}\left(1 - \frac{d}{b} - \frac{e}{c}\right) \quad,$$

where C, the number of seconds per day, is a unit conversion coefficient needed because the time unit for activity is sec, and otherwise the time unit is days.

Bioassay data. We consider a particular case involving a woman with internal exposure, not treated with Prussian Blue (a decorporating agent), 25 years old, 58 kg body weight, and 163 cm height. Intakes may have occurred anytime in the period from September 13 to 24, 1987, and these are coalesced into a single assumed intake on September 20. The bioassay data consist of 15 urine excretion rate measurements (Urine24h) and 6 whole body (W) measurements. The bioassay data are given in Table 1.

Table 1—Bioassay data for an adult female.

Date	Urine(kBq/L)	Urine24h(kBq)[a]	W(MBq)	1 SD(MBq)
04-Oct-87	70.9	85.1		
06-Oct-87	25.2	30.2		
09-Oct-87	17.4	20.9		
10-Oct-87	22.6	27.1		
11-Oct-87	18.5	22.2		
12-Oct-87	10.3	12.3		
16-Oct-87	17.2	20.6		
18-Oct-87	9.44	11.3		
24-Oct-87	9.44	11.3		
06-Nov-87	10.4	12.4		
08-Nov-87	11.5	13.8		
16-Nov-87	4.96	5.95		
24-Nov-87	9.36	11.2		
25-Nov-87			2.74	0.481
27-Nov-87	6.44	7.73		
28-Nov-87	13	15.5		
09-Jan-88			1.82	0.311
19-Jan-88			1.78	0.303
19-Apr-88			0.995	0.126
01-Jul-88			0.555	0.0851
11-Nov-88			0.181	0.0296
[a]**Urine in (kBq/L) multiplied by 1.2 L**				

Prior probability distributions. The prior probability distributions of the parameters were assumed to be uniform in log space from some minimum to a maximum value. The prior for the intake, based on the largest whole body measurement (3 MBq), had a minimum value of 3 MBq and a maximum a factor of 100 larger. For the transfer rates shown in Fig. 1, the minimum and maximum values were, as a starting point, factors of 10 below to 10 above the recommended values (except that the central value for transfer rate 1 in Fig. 1 needed to be changed to 2 d^{-1}; see Fig. 5 below). The recommendations come from a committee of experts and in that sense represent the results of many other experiments studying the behavior of cesium in animals and humans.

MCMC interpretation. Figures 2 and 3 show the data versus the posterior average calculated from a 1000-record tape file with the Metropolis-Rosenbluth-Teller algorithm. Two runs are shown with two extreme starting points and different random number seeds.

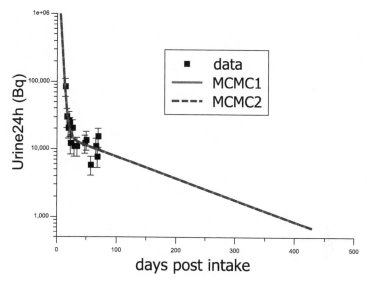

Figure 2—Urine data versus posterior average.

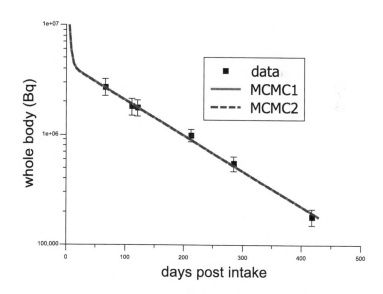

Figure 3—Whole body data versus posterior average.

The urine data was assumed to have a lognormal likelihood, and the GSD (S) value was chosen to be 1.35 ($S = 0.3$) in order to have $\langle \chi^2(\theta) \rangle = 1$ per data point. This is a reasonable value for GSD, based on knowledge of variability of urinary excretion for volume-normalized samples.

The MCMC calculations used the Metropolis-Rosenbluth-Teller and the Multiple-Candidate algorithms, with parameters as shown in Table 1.

Table 1 – Comparison of performance of MRT and 100-candidate MC algorithms. The quantity Δ is the random-walk halfstep, Δ_I referring to intake and Δ_k referring to the transfer rates. With efficient parallel processing, the MC algorithm would be a factor of 100 faster than the time shown.

algorithm	Δ_I	Δ_k	$N_{iterations}$	N_{moves}	time(min)
MRT	0.05	0.5	4×10^7	1.6×10^6	1.4
MC-100	1	1	10^6	2.5×10^5	3.4

The MCMC calculations used grouping of parameters, with 6 groups that were moved in random order, each group consisting of one of the transfer rates from Fig. 1 plus the intake. The total number of chain iterations required for convergence (of the dose) is shown as well as the total number of times the chain moved. The acceptance fraction is the number of moves divided by the total number of iterations (4% for MRT and 25% for MC). For both runs the first 10% of the run was disregarded as being influenced by initial transients.

From the standpoint of the dose, the most important transfer rates are the smallest, because they represent the bottlenecks in the elimination of the material from the body. Figure 4 shows a scatter plot of k_6 versus k_5 defined in Fig. 1 using the Metropolis-Rosenbluth-Teller algorithm. Two runs are shown with two extreme starting points and different random number seeds.

186

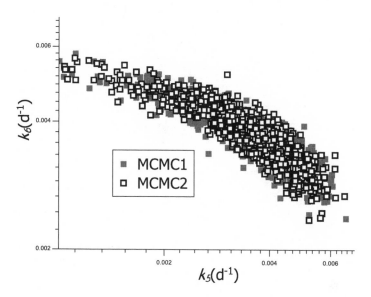

Figure 4—Scatter plot showing the distribution and correlation of the two smallest transfer rates.

Similarly, Fig. 5 shows a scatter plot of intake versus k_1 .

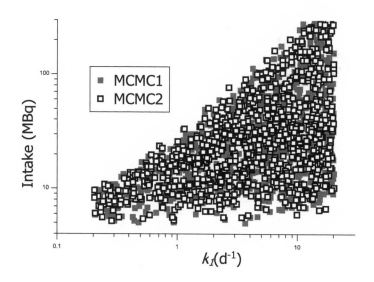

Figure 5—Scatter plot of intake versus transfer rate k_1.

Figure 6 shows the bottom-line dose quantity, the number of nuclear transformations.

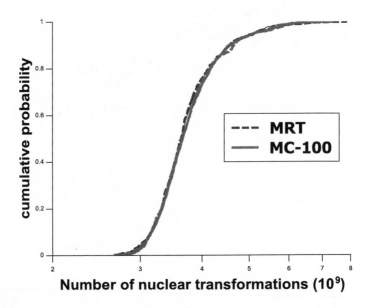

Figure 6—*Distribution of the number of nuclear transformations, representing dose, calculated using the Metropolis-Rosenbluth-Teller and MC-100 algorithms.*

The number of transformations is $3.8 \pm 0.7 \times 10^9$ (average $\pm$ standard deviation), which corresponds to a rather small radiation dose, not of health concern.

The distributions shown in Figs. 2 through 6 are obtained from a tape file of 1000 iterations evenly spaced throughout the entire run. However, even a 100 record tape file is effective in summarizing the result of the calculation. From this tape file a distribution of any desired function of the parameters can be calculated without rerunning the chain.

From Fig. 5 one sees that the intake is not limited, except by the prior. This is because the minimum of χ^2 is extremely broad in terms of intake (it is actually constant above a threshold value of intake, see Exercise 1 of Chapter 17). This leads to the high dose portion of Fig. 6, which involves large intakes. Measurements of fecal excretion would rule out these large intakes. Exploring the entire range of intakes takes many chain iterations for convergence.

Note that the MC-100 algorithm converges with 40 times fewer chain iterations (but 2.5 times more forward calculations). With efficient parallel processing the time for MC would be 40 times less than MRT.

It is interesting that the central parameter values obtained for this case are somewhat different from the committee recommendations shown in Fig. 1. Of course, a committee cannot predict with any certainty the metabolism for any individual person. Going forward, it would make sense for expert consensus like that of the ICRP to be expressed as prior probability distributions rather than point values of parameters, or as discrete collections of models, as will be discussed in the next chapters.

Exercises

1. Using the Fortran program CS137 contained with the supplementary material for this chapter in the subfolder CS137, repeat the MCMC calculations for the MRT algorithm with Δ_I for intake equal to 1 rather than 0.05. What happens?
2. Explain the correlation shown in Fig. 4.

Chapter 17. Example—Discretization of Continuous Variables

This example reconsiders the Goiânia example of the previous chapter using a discretization technique where the model transfer rates rather than being continuous variables are discretized into a set of $l = 1,...L$ discrete values. This technique is useful when the forward model calculations take a long time such that the numbers of forward calculations necessary for straightforward MCMC is prohibitive. This technique lends itself to parallelization because the L model calculations are independent and can be done simultaneously beforehand and the results stored in interpolation tables for later use.

For the Goiânia problem at hand, the intake amount I is allowed to remain a continuous variable. The biokinetics are a linear function of intake amount that can be recorded as linear interpolation tables of the model quantities (in this case B, S, T) versus time t for unit intake. Initially we calculate L biokinetic model interpolation tables. Then we perform the calculation of the posterior probability $P(I, l \mid data)$ of the intake amount I and the biokinetic index l given the measurement results. As usual, MCMC calculates a tape file containing a sample from the posterior distribution of (I, l). The same prior used in Chapter 16 is used here, where there is a factor of 100 possible variation of the transfer rates and the intake amount. The total number of biokinetic models was $L = 100000$, which was about the maximum for the machine used. Increasing L by a factor of 10 caused an allocation error, because of running out of RAM.

Three methods of discretization. The discretization process can be done in different ways, such as random sampling, regular sampling, or stratified-random sampling.

1) Using random sampling, the L sets of model parameters are generated from the prior distribution discussed in Chapter 16 ($i = 1,...6$ independent log-scale uniform distributions) as follows:

$$\log k_{i,l} = \log k_{i,\min} + x_l \left(\log k_{i,\max} - \log k_{i,\min} \right) \quad ,$$

where x_l is a random number uniformly distributed between 0 and 1, and $k_{i,\min}$ and $k_{i,\max}$ are the minimum and maximum values of parameter k_i from the prior. For each of these L sets of 6 parameter values, a biokinetic model interpolation table is calculated and stored.

2) In regular sampling, each parameter is divided into some number of slices, and the parameter value is the center of the slice. Referring to Fig. 4 of the previous Chapter, the extent of the modal region for transfer rate k_6 is about 0.003 to 0.004 or a fraction

$$\log(0.004/0.003)/\log(100) = 0.062 \cong 1/16$$

of the total extent of the prior (a factor of 100). Similarly, let us say the modal extent for transfer rate k_5 is about $1/10$. If the number of slices for these two parameters are 10 and 16, with regular sampling, the 6 parameters might be sliced up in this way (giving 100000 total):

$$5 \times 5 \times 5 \times 5 \times 10 \times 16 = 100000 .$$

With regular sampling perhaps 1 of these 100000 points would be in the modal region in a k_5 vs k_6 plot usch as Fig. 4 of Chapter 16.

3) However, with stratified-random sampling, where the parameter is chosen randomly within the slice and with 100000 total samples, the number of different values of (k_5, k_6) in the modal region will be $100000/160 = 625$ rather than 1. The same applies on average for random sampling.

Stratified-random sampling would seem to offer the advantage of continuity. Each of the 6 sub indices (running from 1,5 ;...; 1,16) could have it's own random walk half step. This would seem to offer the advantage of continuity, because the forward model is a continuous function of the parameters mapped from the sub indices. This scheme would be more complex, and, thankfully, it is not necessary. Random sampling using a one-dimensional index $l = 1,...L$ offering no continuity of forward model versus index l is found to work.

Results using random sampling. Performance using random sampling for two mixed-Δ algorithms (50% of time $\Delta = 1$) is shown in Table 1. The MCMC calculations had two parameters, the intake I and the biokinetic model index l, with no grouping of parameters (both parameters moved at every iteration).

Table 1 – Comparison of performance of MRT and 40000-candidate MC algorithms. The quantity Δ is the random-walk halfstep. With efficient parallel processing, the MC algorithm would be a factor of 40000 faster than the time shown.

algorithm	Δ_I	Δ_l	$N_{iterations}$	N_{moves}	time(min)
MRT	0.01	1	4×10^8	23000	4.9
MC-40000	0.03	1	10000	5000	4

Figures 1 and 2 show the data versus the posterior average calculated using MCMC from a 1000-record tape file. Two runs using are shown with two extreme starting points and different random number seeds.

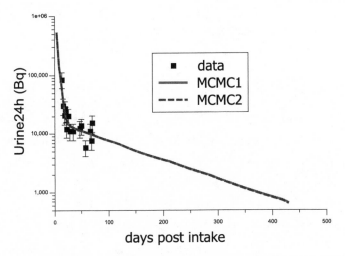

Figure 1—Urine data versus posterior average.

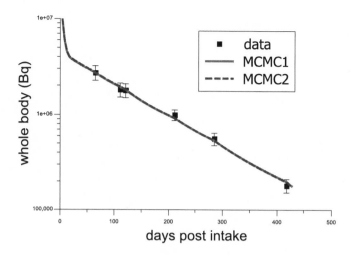

Figure 2—*Whole body data versus posterior average.*

Figure 3 shows a scatter plot of k_6 versus k_5 using the MRT algorithm. One sees that this plot is very similar to Fig. 4 of the previous Chapter.

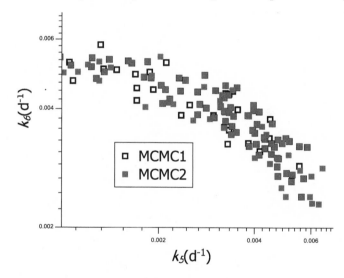

Figure 3—*Scatter plot showing the distribution and correlation of the two smallest transfer rates.*

Similarly, Fig. 4 shows a scatter plot of intake versus k_1, which is also very similar to Fig. 5 of the previous Chapter.

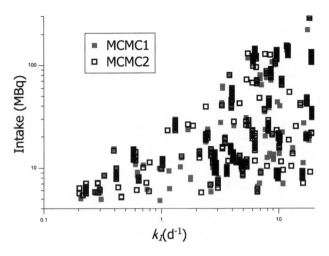

Figure 4—*Scatter plot of intake versus transfer rate* k_I.

Figure 5 shows the bottom-line dose quantity, the number of nuclear transformations, which determines the radiation dose to the person.

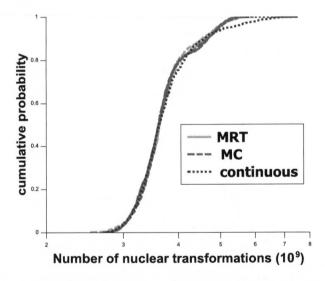

Figure 5—*Distribution of the number of nuclear transformations, representing dose, calculated using the MC and MRT algorithms. Also shown is the continuous calculation of the previous Chapter.*

There is very good agreement of the MRT and MC algorithm results, and the calculation of the previous chapter using continuous variables is quite close. The number of transformations is seen to be about 3.6×10^9.

194

Figure 6 shows the distributions of the number of nuclear transformations as above compared with two cases having $L = 1$.

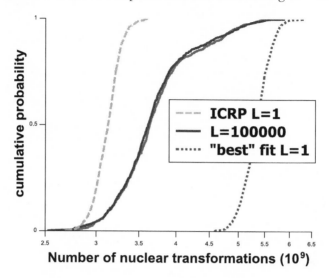

Figure 6—Distribution of the number of nuclear transformations compared to two $L = 1$ cases, one using ICRP- recommended values of the transfer rates and the other from a minimum χ^2 fit determining the transfer rates.

The two $L = 1$ cases are 1) using the transfer rates from Fig. 1 of Chapter 16–the ICRP "recommended" values, and 2) the transfer rates obtained from a minimum- χ^2 fit using the GN algorithm discussed in Chapter 12 (see also, Exercise 1). The average χ^2 values are shown in Table 2.

Table 2 – Values of $\langle \chi^2 \rangle$ per data point and dose for the cases shown in Fig. 6.

case	$\langle \chi^2 \rangle / m$	Dose (10^9)
"Best" fit $L = 1$	0.7	5.3 ± 0.3
$L = 100000$	1	3.6 ± 0.5
ICRP $L = 1$	3	3.1 ± 0.2

From Fig. 6 one sees that the two $L = 1$ cases, while included in the prior, have very small posterior probability and seem to represent low and high dose extremes. From Table 2 one sees that having variability of the transfer coefficients has only about a factor of 2 effect on dose uncertainty.

If L is not large, because of the coarseness of the biokinetics grid, the posterior probability will be concentrated on a single value of l, the one giving the smallest χ^2.

In practice, using this technique, one looks at the final value of $\langle \chi^2 \rangle / m \cong 1$. If $\langle \chi^2 \rangle / m$ is significantly larger than 1, the collection of biokinetic models needs to to be expanded.

Exercises

1. Using the Fortran program Cs137GN contained with the supplementary material for this chapter in the subfolder Cs137\GN, calculate the minimum χ^2 values as a function of the intake amount, showing that a constant minimum value is attained above a threshold value of intake.

2. Explain why the quantity Δ_l in Table 1 should make very little difference.

Chapter 18. Code Validation Example from Internal Dosimetry

An important tool in probabilistic data modeling is the numerical simulation of random variables using a computer. A data analysis program designed to handle a certain type of data can then be tested by using computer-generated data, for which the correct answer is known. This is done several times in this book. It is particularly important outside of pure scientific research, where the calculations are interfacing with "civil life," to do this with some formality, because these code validation examples are a recognized and understandable part of software quality assurance.

This example, again from the field of internal dosimetry, is important because it cannot be solved in a non-probabilistic way, making use as it does of a very informative prior (the alpha prior). Also, the number of parameters is greater than the number of data points.

The example is realistic, and the data, although numerically simulated, are very similar to real data. The measurements are counting measurements with small numbers of counts. A worker is monitored for exposure to the radioactive material ^{238}Pu. This material, which has a half-life of 87.7 years, is used as a heat source, one gram generating approximately 0.5 watts of power. It is an alpha emitter and a significant health concern, but only when taken inside the body. The assumed route of intake in this case is inhalation.

Discretized forward models. In Chapter 16 we have seen an example of a biokinetic model. These models are developed by studies of animal and human data. All the biokinetic models used in radiation protection are, so far, linear in the amount of intake and are compartmental models of the type of the cesium model. However, the biokinetic models for plutonium are much more complicated than that for cesium. Also, the dose calculation is generally more complex than for cesium, which is uniformly distributed throughout the body. The material might preferentially seek out certain body tissues. For plutonium these are the liver and bone surfaces, and therefore the body is not being uniformly irradiated. Notwithstanding all this complexity, because of linearity with amount of intake, the model is summarized by an interpolation table giving the bioassay and dose quantities of interest per unit intake amount as functions of time.

We use the discretization techniques discussed in Chapter 17 where a discrete collection of biokinetic models is used, each model having a particular set of parameters. This collection of models constitutes the biokinetic prior, that is, the "universe of possibilities" for the biokinetics. When the discrete collection is very large the discrete and continuous approaches are the same. With a discrete collection of biokinetic models, the multidimensional integration of Chapter 16 becomes summation over the models and one-dimensional integration over the intake amount, which can be done either with MCMC or with one-dimensional numerical integration and summation. Now in this chapter, instead of only a single intake, we are allowing many intakes. This requires MCMC. For simplicity and to correspond to existing practice in the field of internal dosimetry, the discrete collection of biokinetic models is small, but large enough to obtain $\left\langle \chi^2(\theta) \right\rangle / m \cong 1$.

As in the previous example, the experts involved with the International Commission on Radiation Protection (ICRP) play a role in this. Based on ICRP recommendations, one can assemble a discrete collection of models that constitute a biokinetics prior for plutonium 238. We assume 8 possible models for inhalation. These start with type S or type M (Slow or Medium dissolution in lung fluids according to the ICRP designation) with three different particle sizes. This basic collection is augmented by two additional models with the biokinetic behaviors actually observed in the Los Alamos workplace, but outside the scope of the ICRP recommendataions. The prior on biokinetic behavior consists of this discrete collection of 8 models with equal probability. In other situations one might want the model that the ICRP designates as the default model (recommended for use in lieu of other information) to have a higher prior probability.

Simulated data. Twice-yearly urine bioassay samples are taken from the worker. These samples are analyzed for the presence of [238]Pu using chemical separation of the plutonium followed by alpha spectrometry (detection of the energy of the emitted alpha particles), using 20-hour counts.

198

The simulated urine data for one worker over a period of ten years are shown in Table 1 below.

Table 1—^{238}Pu bioassay measurements. In the 20-hour counting period T, N sample counts are detected, while N_B background counts are detected in a counting period 6 times as long. Because of an informative prior on background, the background gamma distribution has $\alpha_B = N_B + \alpha_0$ and $\beta_B = 6T + \beta_0$ where the background prior has $\alpha_0 = 2$ and $\beta_0 = 6T$. The quantity A is the normalizing coefficient, converting the bioassay quantity into counts.

sample date	N	N_B	A (count/mBq)
2-Aug-1996	0	3	25.58
1-Feb-1997	0	4	8.20
2-Aug-1997	4	8	25.13
1-Feb-1998	2	3	18.42
3-Aug-1998	3	1	22.47
1-Feb-1999	3	2	17.01
3-Aug-1999	4	3	16.08
2-Feb-2000	2	1	18.80
2-Aug-2000	5	0	12.14
1-Feb-2001	4	1	14.51
2-Aug-2001	2	0	15.46
1-Feb-2002	2	2	10.79
3-Aug-2002	0	0	12.52
1-Feb-2003	0	3	13.53
3-Aug-2003	4	1	12.18
2-Feb-2004	1	3	15.15
2-Aug-2004	0	1	14.37
1-Feb-2005	0	4	15.17
2-Aug-2005	2	0	17.27
1-Feb-2006	2	3	14.77

The simulated data for one test case are calculated as follows. First, a biokinetic model is chosen randomly from the prior collection. For this one test case, the model was ICRP type S and 1 μ m Activity Median Aerodynamic Diameter (AMAD) particle size. The true value of 24-hour urine excretion is calculated assuming a single intake occurred on 15-Mar-1997. The amount of intake corresponds to a 50-year effective dose to the whole body (CED) of 5 mSv, which is a small dose at the threshold of regulatory concern. The biokinetic model is used to simulate the true value of the urine excretion ψ at any time after the intake.

To simulate the observed number of counts, the true mean value of the number of counts is assumed to be given by

$$\mu = A\psi + B \quad , \tag{1}$$

where ψ is the true value of 24h urine excretion from the biokinetic model, A is the normalizing normalizing coefficient, and B is the background. The background has been discussed in Chapter 4. The background measurement is assumed to be for $T_B / T = 6$ sample counting periods. The true value of background counting rate λ_B was generated from its prior distribution, which was assumed to be a gamma distribution with $\alpha_0 = 2$ and $\beta_0 / T = 6$ (that is an average of 1 count in 6 counting periods). Then, the number of background counts N_B was generated from a Poisson distribution with $\mu_B = \lambda_B T_B$, simulating the background count measurement. Given this number N_B, the background B in Eq. (1) above was from the posterior distribution; that is, it came from a gamma distribution with $\alpha_B = N_B + \alpha_0$ and $\beta_B = T_B + \beta_0$. The background scaling factor assumed for Table 1 is $R = \beta_B / T = 12$.

The nominal (median) value of the normalizing coefficient A in Table 1 is generated from a lognormal distribution with median = 14 count/mBq and $S = 0.3$ in order to simulate variations in chemical recovery.

The lognormal uncertainty of A in Eq. (1) was also assumed to be given by $S = 0.3$, corresponding to the urine collection protocol used.

The true value of the Poisson mean number of counts was calculated using Eq. (1). The number of counts appearing in Table 1 was then generated from a Poisson distribution.

The exact likelihood was calculated using the data in Table 1. Each data point is then represented by an interpolation table of values of the normalized residual $\chi(\psi)$ (the number of standard deviations the data departs from the true value) versus the calculated true value of the urine excretion ψ.

A large collection of test datasets are generated so that the program being tested can be assigned a "score," which is a fraction of cases that are "correct" by some criterion.

Probability distribution of intake amount. In real life it is known from experience that very few intakes occur, and the prior on intake amount was given by the alpha prior with $\alpha = 0.001$ per year, corresponding to a very small probability of intake in a year. The actual amount of intake ξ is obtained from the theta variable generated uniformly from 0 to 1 using the formula,

$$\xi = \xi_{max}\,\theta^{1/(\alpha\Delta t)} \quad , \tag{2}$$

where ξ_{max} is the maximum possible intake (chosen to be larger than any conceivable intake), and Δt is the 1-year interval. The power in Eq. (2) in this case is 1000, which means that the prior on intake amount is very concentrated at 0 intake, and multiple intakes are very unlikely.

In the modeling or interpretation of this simulated data, intakes were assumed possible in any of the 11 years from the beginning of the data in 1996 to its end in 2006. The time of intake was variable within each year-long interval, and the biokinetic type was also variable. There were therefore 11 intake amounts and 33 parameters total for 19 data points, excluding the first measurement, which defines the start of work.

MCMC calculations. In the MCMC calculation the parameters were assigned to groups corresponding to each intake. The parameters in one group were the three θ values giving intake amount, time of intake, and biokinetic type, which is an integer obtained from its corresponding θ values by a stepwise-constant function corresponding to the prior probabilities of the biokinetic types. At each iteration, the chain focused attention on only one group, and the Metropolis-Rosenbluth-Teller algorithm was used with a single candidate, requiring about 2×10^6 chain iterations before convergence of year-by-year effective dose was reached. The computer time required was about 2 seconds. The MC algorithm can also be used, but without parallel processing requires somewhat more computer time.

The convergence criterion considered the results of two maximally different MCMC runs (1 and 2) with different starting points (all θ's 0 and all θ's 1) and different random number seeds done in parallel, with the requirement for convergence being that

$$SD_{12} < \frac{1}{3} Avg_{12}(SD_{MCMC}) \quad ,$$

that is, the standard deviation of the average of the quantity calculated from MCMC runs 1 and 2 ($SD_{12}(x) = (x_1 - x_2)/\sqrt{2}$) is less than 1/3 of the average over runs 1 and 2 of the MCMC standard deviation of the quantity. The quantities considered for convergence were the year-by-year $E(50)$ doses corresponding to intakes in a particular year, and all had to satisfy the convergence criterion. The basic idea of this criterion is to ensure that the difference between the average values for the two runs of the quantity of interest is small in relation to the intrinsic uncertainty of the quantity.

The "result" of the MCMC run was a sample of 1000 parameter values evenly spaced throughout the entire run and written on a "tape" file. These constitute, after eliminating the "burn in" iterations, 900 alternate possible interpretations of the data. From this tape file a distribution can be made of any function of the parameters, and thus, using the forward model, 900-sample distributions of any desired quantity can be quickly calculated.

Data versus posterior mean results are shown in Fig. 1. The two curves show the two runs MCMC1 and MCMC2 with maximally different starting points and different random number seeds.

202

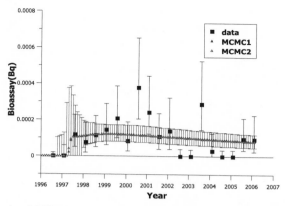

Figure 1—MCMC internal dosimetry calculation using simulated data.

Figure 1 shows the data, graphically represented as the maximum of the exact likelihood function, with the distance to the upper and lower $\chi(\psi) = \pm 1$ points as the error bars. The curve also shows the 5% and 95% limits of the posterior average from the forward calculation. The posterior average $\langle \chi^2(\theta) \rangle / m$ is very close to 1.

The cumulative posterior distribution of 50-year committed dose for intakes in 1996, 1997, and for all years is shown in Fig. 2.

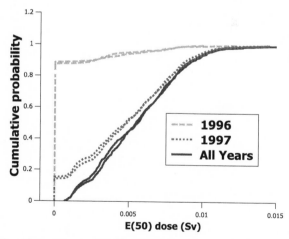

Figure 2—Cumulative probability of the year-by-year and total E(50) dose.

As usual the double curves show the results with different extreme starting points (all θ's either 0 or 1) and different random number seeds.

The median of the total $E(50)$ dose is very close to the true value of 0.005 Sv. The distribution of the total $E(50)$ dose for all years departs from 0 meaning that there is very high probability that some intake or intakes have occurred. Because of the very small probability of intakes from the alpha prior, a single significant intake is wandering around slightly from 1996 to 1997.

Notice how the alpha prior plays an important role in the interpretation of this data. If, instead of Eq. (2), some more uniform distribution of potential intakes were used, the "fit" shown in Fig. 1 would have many more small steps, indicating multiple intakes, in response to random variations of the data.

Figure 3 shows the year-by-year $E(50)$ dose, with the uncertainty bars representing the 5% and 95% points around the posterior median.

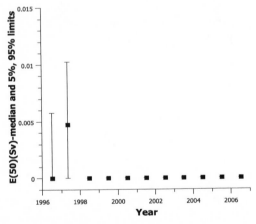

Figure 3—Year-by-year 50-year effective whole body dose $E(50)$.

Exercise

1. Using the Fortran program ID.FOR (batch file RUN.BAT) contained with the supplementary material for this Chapter, rerun the example discussed in this Chapter.

Chapter 19. Real-Life Internal Dosimetry Example

This chapter is based on a real-life example from internal dosimetry. A worker has submitted urine bioassay samples for exposure to plutonium 238, and the requirement is to determine the worker's radiation dose. The dose quantity of interest is the 50-year effective whole-body dose by year of intake, or committed effective dose, denoted by $E(50)$.

Specifying likelihood functions using actual recorded measurement quantities. In this book the attempt is made to treat the likelihood function realistically in terms of the actual measurements. Unfortunately, sometimes complete information about the measurement is not recorded. The example given in this chapter is such a case, where we know the measurements were counting measurements, but only the "result" and it's standard deviation were recorded, without the actual count quantities. The count quantities, as we have seen in Chapter 7, are A, the median value of the normalizing coefficient converting net counts to bioassay units, N and N_B, the sample and background counts, and R, the background scaling factor (usually the ratio of background count time to sample count time). Using the formula for the result y from the count quantities,

$$y_{meas} = \left(N - N_B / R\right) / A \quad .$$

In Chapter 7 the normalizing coefficient was discussed. It is given by the formula

$$A = \Delta t_x \varepsilon_r \varepsilon_c \Delta t_c \quad ,$$

where, for a 24-hour urine sample, the excretion time is Δt_x (in units of 24 hours), the efficiency of the chemical separation (chemical recovery fraction) is ε_r, the counting efficiency is ε_c, and the count time is Δt_c. For a 24-hour sample, $\Delta t_x = 1$, and the efficiencies can be estimated, so A depends mostly on the count time, which is usually known roughly in situations where long counts were necessary. Also, the background counts and count time ratio can be reasonably guessed. In this way, by estimating A, N_B, and R, one obtains

$$N = Ay_{meas} + \frac{N_B}{R} \quad .$$

The background counts may be adjusted to insure that the sample counts are never negative (when y_{meas} is negative).

Another way of reconstructing the count quantities when the measurement standard deviation σ_{meas} is available, is to determine or guess the formulas used to calculate y_{meas} and σ_{meas} from the count quantities and assume values for N_B and R. Then, write the formula for the measurement result divided by it's standard deviation for given N_B and R, which yields a quadratic equation that may be solved for N. If it is truly known what formulas were used for y_{meas} and σ_{meas}, a refinement is to step through integer values of N_B until the solution for N is close to being integer.

Instead of using back-calculated count quantities, the normal approximation version of Eq. (9) of Chapter 7 can be used to obtain the Likelihood directly as a function of y_{meas}, σ_{meas} and S_{norm}, which is the approach that we will use here. The use of count quantities is not so critical when the data are mostly positive and large in terms of standard deviations.

For the case under consideration, the bioassay data are as shown in Table 1 below.

Table 1—^{238}Pu urine bioassay measurements for worker. Corresponding to the urine collection protocol used (specific-gravity normalization to 24-hour collection), in addition to the measurement uncertainty Standard Deviation σ as shown, there is a lognormal normalization uncertainty estimated to be $S_{norm} = 0.3$ (GSD$=1.35$).

Count	Date	Urine24H(mBq)	σ (mBq)
1	25-Oct-1978	-0.185	0.37
2	18-Dec-1978	-0.333	0.37
3	5-Apr-1979	0	0.37
4	20-Jul-1979	-0.888	0.37
5	18-Oct-1979	0	0.37
6	16-Jan-1980	-0.222	0.37
7	31-Mar-1980	-0.111	0.37

8	23-Jun-1980	0.222	0.37
9	7-Sep-1980	-0.148	0.37
10	3-Nov-1980	0.777	0.37
11	13-Nov-1980	0.481	0.37
12	23-Dec-1980	1.591	0.3811
13	22-Jan-1981	1.813	0.4181
14	18-Mar-1981	0.74	0.37
15	21-May-1981	1.776	0.4107
16	23-Jun-1981	6.623	1.0064
17	27-Jul-1981	10.471	1.3764
18	14-Sep-1981	4.773	0.8066
19	21-Sep-1981	0.555	0.37
20	26-Oct-1981	0	0.37
21	3-Dec-1981	4.736	0.8029
22	24-Feb-1982	2.294	0.4921
23	28-May-1982	3.441	0.6475
24	21-Nov-1982	4.884	0.8214
25	22-May-1983	3.552	0.6586
26	15-Nov-1983	1.887	0.4292
27	22-May-1984	3.7	0.6771
28	2-Nov-1984	8.621	1.2062
29	10-May-1985	0	0.37
30	31-Jul-1985	-0.148	0.37
31	4-Nov-1985	3.811	0.6919
32	30-Apr-1986	0	0.37
33	11-Nov-1986	5.55	0.8954
34	1-Jul-1987	4.551	0.7807
35	22-Oct-1987	11.433	1.4615
36	6-Jun-1988	8.325	1.1766
37	21-Nov-1988	4.699	0.7992
38	13-Apr-1989	6.179	0.962
39	16-Nov-1989	4.144	0.7326
40	11-May-1990	3.7	0.6771
41	14-Jul-1990	5.106	0.8436
42	23-Mar-1991	7.178	1.0656
43	29-Mar-1991	4.81	0.8103
44	14-May-1991	6.364	0.9805
45	14-Jan-1992	3.182	0.3145
46	2-Jul-1992	5.735	0.74
47	25-Apr-1994	3.108	0.629
48	14-Nov-1994	2.923	0.814

In this case it is known that the worker was involved in an "incident" on 31-Oct-1980. On that date some sort of accident occurred that might have resulted in an inhalation intake by the worker. These incidents usually involved the detection of measurable external contamination or an air monitor alarm. Therefore, in the intake year including this date (1980), an intake with fixed date of intake was added with a broad lognormal prior on intake amount used in addition to a variable date of intake and the alpha prior on intake amount, which is used in all years.

MCMC interpretations. The MCMC calculation is shown in Fig. 1. As usual the two curves show runs with maximally different starting points and different random number seeds. In addition to the data likelihood functions, the plots show the 5% and 95 % limits of the posterior of the calculated bioassay quantity.

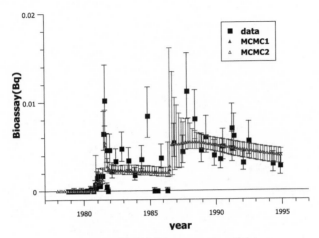

Figure 1—MCMC interpretation of data.

The quantity $\langle \chi^2(\theta) \rangle / m = 3$ is large. Looking at Fig. 1 and the data in Table 1, we see that 5 data points seem anomalously low, outside of reasonable bounds of biological variability of urine excretion. If we eliminate these 5 data points, $\langle \chi^2(\theta) \rangle / m$ decreases to 1.5, and we obtain the result shown in Fig. 2.

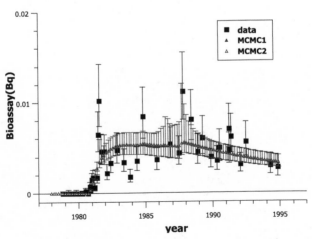

Figure 2— MCMC interpretation of data after eliminating 5 low data points.

Figure 3 shows the corresponding year-by-year $E(50)$ dose.

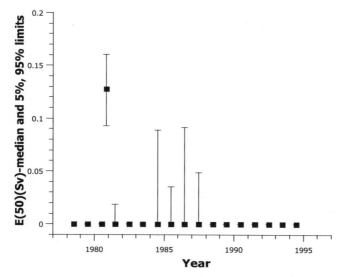

Figure 3—Year-by-year 50-year effective whole body dose $E(50)$.

The data are consistent with a single intake, associated with the incident, occurring in 1980.

A collection of 8 different biokinetic models formed the biokinetic prior as discussed in Chapter 18. It is interesting that the posterior probability of biokinetic type for the 1980 intake is 100% of a type not included among the International Commission on Radiation Protection recommendations. However, it is a type of biokinetics previously seen in this workplace and added to the prior collection on the basis of this experience. This type is distinctive because there is a slow onset of urine excretion following the intake.

Exercise

1. Using the Fortran program ID.FOR (batch file RUN.BAT) contained with the supplementary material for this Chapter, rerun the example discussed in this Chapter.

Chapter 20. Afterword—Probability and Science from Laplace to Feynman

With the work of this little volume complete, I ask the reader to allow me a small space to philosophize about probability and science and other possible ways to think about probability.

Probability is the mathematical language needed for science, and all scientific measurements, if taken to the most precise level, are variable and unpredictable to some degree.

There is a certain lack of elegance in the mathematical expression of probability, for example, in the Markov Chain theory. If the probabilities are positive, it would be nicer if this were a natural consequence of the mathematics, if, for example, the probabilities were expressed as the square of something or the exponential of something or by some other means.

Electrons passing through two slits. Moving forward several centuries from the time of Laplace, let us consider the physical example illustrated in Fig. 1.

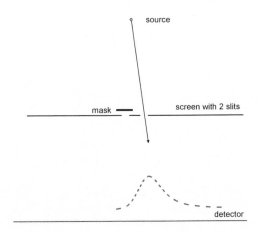

Figure 1—A point source of electrons above a detector with a screen having two slits in between.

210

Many electrons are emitted, and some pass through the single uncovered slit and are detected. The detector shows the position of the detected electron.

What would Laplace say? Laplace would look at this data and (after being brought up to date about an electron being a small charged particle that is a fundamental constituent of matter) perhaps wonder why the slit in the screen doesn't cast a sharp shadow. But he would no doubt say that if the other slit were opened, allowing an electron to arrive at the detector through one slit or the other, the probability of the electron arriving at a particular point on the detector would have to be the sum of the probabilities for the left slit and the right slit. Laplace would even calculate the probability of the "cause" (passage through left slit or right slit) given the observed position, and electrons detected toward the right would have a greater probability of passage through the right slit.

The truly remarkable thing, which would surely amaze and delight Laplace, is that if one actually were to do this experiment, a result like that shown in Fig. 2 would be obtained.

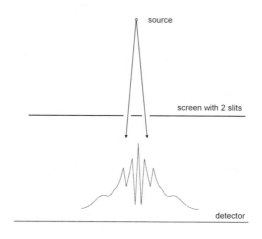

Figure 2—A point source of electrons above detector with a screen having two slits in between.

With two slits open, at some positions there is less probability than with one slit!

From the perspective of modern physics, one might say that the electron can go through one slit or the other, but the probabilities aren't the thing that is added. What is added are the quantum mechanical amplitudes $e^{iS/\hbar}$, where S is the action and $\hbar$ is Planck's constant, for all the various paths the electron can take from source to detector. These amplitudes are complex numbers that can add and subtract, sometimes partially or even completely cancelling each other out.

This is very nice. Probabilities are positive because they are the square of the absolute value of a complex number! The complex amplitude is the thing that is additive across all possibilities. And who would have guessed without doing the experiments—all the multitude of real, difficult, and messy experiments contributed to by many, many people to lead us to this new place of understanding.

Exercise

1. Allowing the condition $S/\hbar = 2\pi$ to define the quantum mechanical wavelength of an electron, estimate the order of magnitude of the slit separation shown in Fig. 2. How is electron diffraction actually observed?

Reference

1. Brown, Laurie M., Ed. *Feynman's Thesis—A New Approach to Quantum Theory*. World Scientific, 2005. Richard Feynman's 1942 PhD dissertation, "The Principle of Least Action in Quantum Mechanics," and Paul Dirac's seminal 1933 paper "The Lagrangian in Quantum Mechanics."